Physico-Chemical Behaviour of Beneficiated Indian Graphite

Physico-Chemical Behaviour of Beneficiated Indian Graphite

Dr. Anup Kumar Bhattacharya

ALLIED PUBLISHERS PVT. LTD.

New Delhi • Mumbai • Kolkata • Chennai • Bangalore • Hyderabad

ALLIED PUBLISHERS PRIVATE LIMITED

D-5, Sector-2, **Noida**–201 301
Ph. Nos.: 0120-4320295/2542557/4352866 • E-mail: delhi.books@alliedpublishers.com

17 Chittaranjan Avenue, **Kolkata**–700072
Ph.: 033-22129618 • E-mail: cal.books@alliedpublishers.com

15 J.N. Heredia Marg, Ballard Estate, **Mumbai**–400001
Ph.: 022-42126969 • E-mail: mumbai.books@alliedpublishers.com

No. 25/10, Commander-in-Chief Road, Ethiraj Lane (Next to Post Office)
Egmore, **Chennai**–600008
Ph.: 044-28223938 • E-mail: chennai.books@alliedpublishers.com

P.B. No. 9932, No. 15, 3rd Floor (Next to Vijaya Bank), 5th Cross, Gandhinagar, Karnataka, **Bangalore**–560009, Ph.: 080-41530285 / 22386239
• E-mail: bngl.journals@alliedpublishers.com / apsabng@airtelmail.in

Sri Jayalakshmi Nilayam, No. 3-4-510, 3rd Floor (Above More Super Market)
Barkatpura, **Hyderabad**–500027
Ph.: 040-27551811, 040-27551812 • E-mail: hyd.books@alliedpublishers.com

Website: www.alliedpublishers.com

ISBN: 978-93-90951-12-3

Published by Sunil Sachdev and printed by Ravi Sachdev at Allied Publishers Pvt. Ltd., D-5, Sector-2, Noida–201 301

SYNOPSIS

Natural graphite occurs in eruptive rocks and in veins through cracks in sedimentary rocks. Perfect graphite crystals are very rare. Imperfections and grain boundaries present in the graphite are important in determining the properties of the material. Impurities and absorbed gases play an important role in determining the properties of graphite.

The graphite structure has the carbon atoms in a planar array in sp^2 bonding. Thus each layer is bound to its neighbour by Van der Waal's or some other forms of weak binding energy. Any reactivity must be seen as deriving from the crystal edges.

Oxidation reactivity and structure of carbon are interrelated and factors responsible are optical texture, crystallinity, pore structure and catalytic impurities. The former two factors are related to orientation and graphitisability influencing the reactivity of carbon atoms in graphite layers. The third defines the access of oxidants to carbon. Reactivity generally decreases with better orientation and graphitisation.

In India refractory industry is the main user of graphite and consumes about 60% of total graphite for making carbon containing refractory. The total reserves of graphite in India is about 48 lakh tonnes and major graphite suppliers in India are located in three states of India, i.e. Orissa, Jharkhand and Tamilnadu and refractory industry uses graphite from these three states.

Generally at the mine head graphite is recovered from gangue material with F.C. content 20%. After repeated floatation with crushing and grinding, the carbon content increases to 90–95%. Here, we have considered three types of graphite having carbon content 80%, 86% and 93% after beneficiation from three sources for our investigation as these are the qualities which are normally used for manufacture of carbon containing refractories.

Reactivity of carbon is influenced by several factors including (i) origin (ii) pretreatment (iii) intrinsic chemical differences in the carbon (iv) differences in surface area (v) catalysis by impurities and (vi) experimental conditions used. Because of these a universal rate expression on the rate of carbon gasification does not exist. The kinetic steps involved in oxidation of graphite are:

- Transportation of oxygen to the external surface of graphite through boundary layer.

- Diffusion of oxygen into the interior of the pore
- Adsorption of oxygen on the pore surface
- Surface chemical reaction in the adsorbed layer
- Desorption of CO
- Outward diffusion of CO from interior to sample surface
- Transportation of CO from sample surface through boundary layer

To study the behaviour of Indian graphite, the three samples were thoroughly investigated both physically and chemically. Proximate analysis, sp. gravity, specific surface area, particle size were measured. Degree of graphitisation and percent graphitised phases present in the samples were studied by XRD. Nature of impurities present in the samples were studied by XRD as well as through SEM-EDAX. Pores and cracks present in the sample was studied by SEM. Fusion temperature of ash was measured by hot stage microscope and ash analysis was done by XRF. Reactivity of graphite was studied by thermo-gravimetric analysis in oxygen and air atmosphere at three different temperatures i.e. 735°C, 785°C & 830°C.Such temperatures were chosen as it was found that in DTA/TGA of graphite sample in normal atmosphere, maximum weight loss was taking place in the temperature range of 730°C to 800°C.

It has been found that three samples have three different characteristics with respect to physico-chemical properties. Palamau graphite has ash content of 18% with higher amount of alkali, i.e, 3.5%. Degree of graphitisation is 85% and graphitised phase present in it is only 67%. Its sp-gravity is 2.14 and specific surface area is quit high of the order of 1.1 m^2/gm. Ash fusion temperature is 1075°C and impurities are mainly quartz, feldspar, corundum and potassium-iron-magnesium-aluminium-silicate hydrates. Large amount of cracks are present in the structure.

Bolangir graphite has ash content of 12%, degree of graphitisation is 98% and graphitised phase present in it is 87%. Its specific gravity is 2.20 and specific surface area is 0.4 m^2/gm. Ash fusion temperature is 1120°C and impurities are quartz, feldspar, corundum and calcium-magnesium-aluminium-silicate hydrates. Large amount of pores are present in the structure.

Tamilnadu graphite has ash content of 4%, degree of graphitisation is 87% and graphitised phase present in the sample is 93%. Its specific gravity is 2.25 and specific surface area is 1.4 m^2/gm. Ash fusion temperature is 1125°C and impurities are quartz, feldspar, pyrites and alkali-iron-aluminium-silicate hydrates. Large amount of pores are present in the structure.

ACKNOWLEDGEMENTS

The author whole heartedly express his profound sense of gratitude to his supervisor Prof. P. Bandopadhyay, Department of Mineral Engineering Indian School of Mines, for the constant guidance, fruitful suggestions and encouragement throughout the entire course of this thesis work. Thanks to M/s. Parijat, Mining Industries (P) Ltd. & M/s Tamil Nadu Minerals Limited for providing the graphite samples under investigation.

The author expresses deep gratitude to Dr. A. Bhattacharya, R&D Centre, SAIL, Ranchi and his team for allowing the author to conduct experiments in their laboratory days-in and days out. The author acknowledges the help received from Sri D.K. Jain, R&D Centre, SAIL for taking computer runs for curve plotting and report editing.

Author also thanks the management of R&D, SAIL, for their permission to conduct the investigations in R&D laboratory.

Dedications

to

My Parents –
Late Byomkesh Bhattacharya
and
Late Sabita Bhattachary.

Their blessings helped in compiling the whole laboratory work in the form of a book. Hope these findings will help thousands of Indian graphite users in their day to day work.

LIST OF ABBREVIATIONS

ΔH	Heat of transition
ASTM	American society for testing materials
cc	Cubic centimeter
C-C	Carbon-Carbon
BET	Brunuer, Emmet and Teller
XRF	X-ray fluorescence
BSS	British standard sieve
emf	Electro motive force
EDAX	Energy dispersive spectrometry
XRD	X-ray diffraction
g	Degree of graphitisation
LASER	Light amplification by stimulated emission of radiation
α	Fraction of base graphite oxidised and defined by ratio of oxidised graphite to initial graphite
t	Time in seconds
T	Temperature in °C
B.D.	Bulk density in gm/cc
A.P.	Apparent porosity (%)
CCS	Cold crushing strength, Kg/cm^2
HMOR	Hot modulus of rupture, Kg/cm^2
PG, P_o, P_1, P_2	Palamau Graphite
BG, S_o, S_1, S_2	Bolangir Graphite
TG, T_o, T_1, T_2	Tamilnadu graphite
$BG1, BG_1$	Bolangir Graphite of Size (–30+60) BSS
$BG2, BG_2$	Bolangir Graphite of Size (–60+150) BSS
$PG1, PG_1$	Palamau Graphite of Size (–30+60) BSS
$PG2, PG_2$	Palamau Graphite of Size (–60+150) BSS
$TG1, TG_1$	Tamilnadu Graphite of Size (–30+60) BSS
$TG2, TG_2$	Tamilnadu Graphite of Size (–60+150) BSS

LIST OF TABLES

LIST OF FIGURES

CONTENTS

CHAPTER 1

INTRODUCTION

1.1 Introduction of Graphite

Graphite commonly called as plumbago or black lead is a soft, black crystalline form of naturally occurring carbon. Natural graphite occurs generally in large masses but found as nodules in eruptive rocks and in veins through cracks in sedimentary rocks.

The earth crust contains somewhat less than 0.2% carbon, most of it as coal, organic materials, graphite, diamond and carbonate minerals. Since the crust also contains about 50% oxygen, it seems reasonable to assume that the major part of the carbon present in the crust during prehistoric cooling was in the form of carbon dioxide and carbon monoxide, depending on the temperature of equilibrium. Natural graphite is often found in connection with carbonates and other carbon compounds and could be the result of their decomposition under conditions of high pressure and high temperature. Graphite cannot be melted in commercial processes to eliminate the structure, owing to its past history. Perfect crystals of graphite are very rare indeed and the imperfections and grain boundaries present in the materials are as important in determining the properties of material as is the properties of perfect crystal. Impurities and absorbed gases also play an important role.

Carbon materials in use are composed microscopically and macroscopically of a large number of carbon grains. The adhesion of the grains strongly influences the property since its lack brings about defects and the voids. Carbon materials are basically porous, having pores and cracks, of which sizes are distributed from several angstrom units to several millimeter. The cracks often run along the lamellae of the carbon plane and optical texture. The mosaic texture resists crack propagation because the lamellae of the carbon plane changes their orientation at the interface of the mosaic units, whereas crack run along the flow texture. Thus the optical texture is closely related to the mechanical strength. The shrinkage due to densification by the calcination and graphitization is also governed by the optical texture because it influences both the extent and direction of the densification. Table 1.1 summarizes some typical properties of graphite.

Table 1.1: Typical Properties of Graphite

Density (g/m^3)	-	~ 2.26
Thermal Conductivity (cal/cm. sec°C)	-	~4.5
Co-efficient of thermal expansion (×10^{-9}/°C)	-	1~0
Tensile strength psi	-	20–30
Modulus of elasticity (Kg/mm^2)	-	5000–6000
Electric resistivity (M Ω cm)	-	50–60
Hardness (Rockwell)	-	90~100

There are many reports about the relationship between the oxidation reactivity and the structure of the carbon. The influential factors are the optical texture, the crystallinity, the pore structure and catalytic impurities. The former two factors are related to the orientation and the graphitizability respectively, influencing the reactivity of carbon atoms in the graphite layers. The third defines the access of oxidants to carbon. The reactivity of edge carbon atoms in the hexagonal sheet is larger than that of the atoms within the sheet, therefore, its reactivity with the gaseous oxidants generally decreases with better orientation (larger anistropic development) and graphitization. At lower temperature, the chemical reaction of the oxidation is slower than that of the oxidant diffusion. The surface area including the walls of the small pores contributes to the reaction. The rate of reaction competes with that of diffusion at higher temperatures and at further higher temperatures, the diffusion becomes rate determining. Thus, the pores affect the rate of reaction, the smallest pores becoming unimportant to the reaction. Most naturally occurring graphite is polycrystalline. Perfect single crystals greater than 10 micron are quite rare, although they can be prepared with difficulty. Most synthetic graphites are prepared by the high temperature calcinations of pitch/coke blends, etc. They are also polycrystalline. Perfect single crystals of quite large dimensions can be obtained by pyrolytic deposition of carbon from carbonaceous vapours. Under suitable conditions the deposit of carbon can take the form of highly oriented layers. Subsequent treatment of this material can produce quite large single crystals of pure graphite. Such graphite is known as pyrolytic graphite. As per the National Mineral Inventory[(1)] as on 01-04-1990, total recoverable reserves of graphite are placed at 4.8 million tones with an average grade of 10 to 40% fixed carbon, occasionally rising to 60%. The data on world reserves of graphite as on 01-04-95 are given in Table 1.2.

Table 1.2: Recoverable World Reserves of Graphite

Sl. No.	*Place*	*Recoverable Reserves in Lakh Tonnes*
1.0	India (Total)	48
1.1	Andhra Pradesh	2.19
1.2	Jharkhand	3.89
1.3	Karnataka	2.62
1.4	Kerala	5.33
1.5	Maharashtra	8.7
1.6	Orissa	13.5
1.7	Rajasthan	5.01
1.8	Tamilnadu	4.22
1.9	Others	2.1
2.0	China	3100
3.0	South Korea	200
4.0	Mexico	3.1
5.0	Canada	2.7
6.0	Brazil	1.0

In India refractory industry is the main user of graphite and consumes about 60% of total graphite production in making carbon containing refractories for different industries.

Indian refractory industries use graphite from three sources, i.e. Sivagangai district of Tamilnadu, Palamau belt of Jharkhand and Sambalpur/Bolangir district of Orissa. For the present investigation the samples from all these sources have been collected and used. The Palamau graphite used here at latitude of 24.02° and longitude of 84.15° was originally having 15% fixed carbon at the mine head and taken out from the mines in lump form. They are then crushed, and purified by floatation, washing, centrifuging drying, screening process and with repeated floatation for three times fixed carbon content improved to 82% and sample was collected from this lot.

Bolangir graphite was collected from Belpara area of Bolangir district at latitude of 20.35° and longitude of 82.6°. The fixed carbon content of the as received sample from mines is 15–20% and beneficiation is carried put by adoption of the crushing, grinding, floatation, washing, drying and screening process. Repeated floatation is done to improve the percentage of the fixed carbon and pine oil and kerosene is used as floatation medium. In the collected sample, 5 times of floatation was done to achieve the fixed carbon of 87%.

Tamilnadu graphite was collected from the Sivagangai district of Tamilnadu at latitude of 9.8° and longitude of 79.6°. Fixed carbon content of experimental sample was having 20% initially at the mime head. Beneficiation process as mentioned earlier was adopted here also and Otto Kumphu cell is used for floatation process. Six times of floatation done with pine oil and kerosene to achieve a fixed carbon of 94% A detailed flow-sheet of the process adopted for beneficiation of the graphite obtained from mines is shown in Figure 1.1.

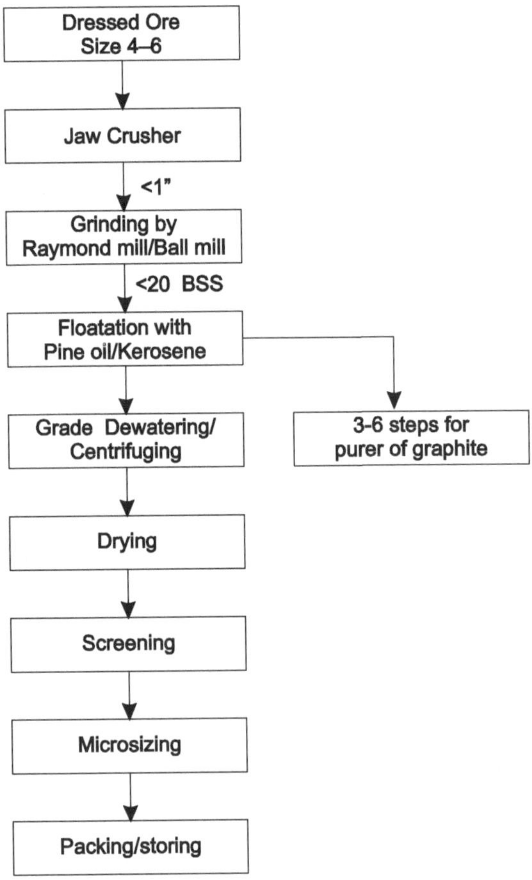

Fig. 1.1: Flow Sheet for Beneficiation of Graphite Ore

CHAPTER 2

LITERATURE REVIEW

2.1 General Features of Graphite

The ideal crystal lattice is the hexagonal α-form, space group D^4_{6h}-C6/mmc. The basal planes consist of open hexagons (interatomic spacing = 0.1415 nm) stacked in an alternating sequence with a lattice spacing of 0.33539 nm.[2]

In practice the lattice constants depends on the graphite source [3] and is also a function of temperature shown in Figure 2.1.

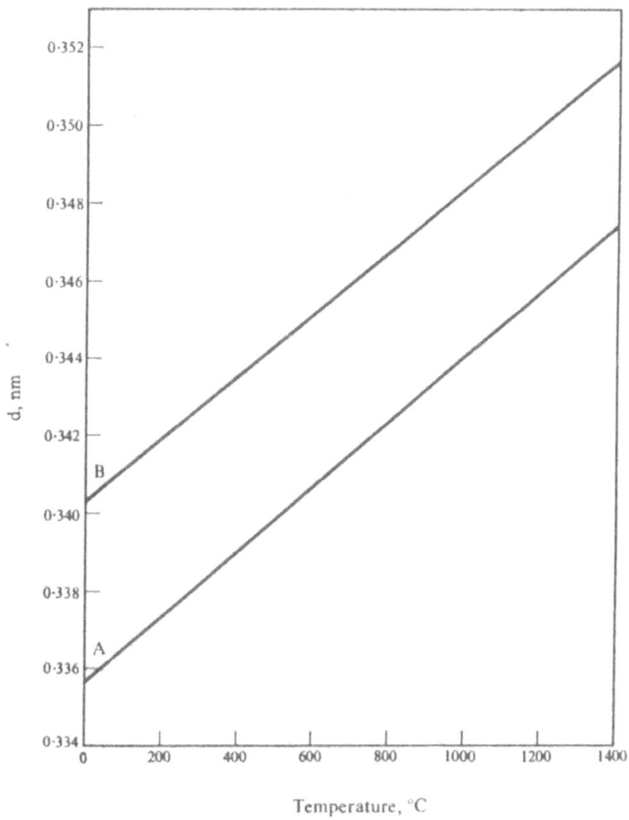

Fig. 2.1: Lattice Constants Dependency on Temperature

The dependence follows the law of $d = do + \alpha T + \beta T^2$ with the constants α and β depending on the type of graphite. For a single crystal of Ceylon graphite.[4]

$$\alpha = 90.54 \times 10^{-6} \text{ and } \beta = 6.33 \times 10^{-6}$$

Each of the structural characteristics of these ideal lattices are subject to modification by defects occurring either in the stacking sequences or in the carbon bond networks. In layer stacking defects, the hexagonal nets may be substantially perfect, but there is no ordered stacking sequences leading to a turbostratic structure. Perhaps, surprisingly the interlayer separation is not very dependent on the degree of stacking order. Bacon [5] has proposed the empirical relation d (A°) = 3.44 – 0.086 (p–1) – 0.064 p (1-p) for the dependence of the interlayer spacing on stacking order 'p'. In practice layer disorder often accompanies breakup of the hexagonal networks and results primarily from edge defects.[6,7] When a C-C bond cannot be formed due to disturbance of the planar sequence the spare carbon valencies may be satisfied in a number of ways. Often foreign atoms can bond to the edge defects[8] or weaker spin paired bonds can be formed with neighbouring atoms.[9] The broken bonds can also lead to holes or ruptures in the structure, the so called "hole and claw" defects.[10] In addition, dislocation in the structure can be detected often associated with the defects. Four major dislocations have been detected, the principle one being defined when both the line and Burgers vectors lie in the basal plane arising from stacking faults.[11] All combinations of edge and coupled edge and screw dislocations have been observed with an average energy of formation of 0.5 erg cm^{-2}. When the line and Burgers vector both lie in non-basal directions only screw dislocations have been observed.

Unit cell of graphite has hexagonal symmetry. The dimensions of the unit cell are[12] a_o = 0.24611 nm and C_o = 0.67078 nm at 288°K. This corresponds to a C-C bond length within the plane of 0.14210 nm indicating a bond order of 1.33.[13] Within the plane each carbon atom, in its sp_2 hybridized state, is bonded to three other carbon atoms. The non-bonded p_z atomic orbitals overlap to give a completely delocalised system of molecular orbitals, thus allowing the transport of electrons throughout the crystal. Hence, graphite is a good conductor. The black colour of graphite also arises from the excitation of these localised electrons. The distance between planes is 0.33539 nm, though this distance is increased slightly in poorly crystalline carbon.[4] This distance is too great for normal valence forces to be operative and successive planes are held together only by weak Van der waal's forces.[14] The virtual lack of bonding between planes leads to ready cleavage along the planes and thus accounts for the softness of the solid. Moreover, because of the lack of interaction between planes it is possible to insert other chemical groups into the crystal, giving rise to the lamellar chemistry of graphite. In fact, the lubricating properties of graphite probably arise as a result of the bonding of oxygen or organic molecules to the graphite layers, thus reducing still further the interaction between planes.

The ideal graphite crystal[15,16,17,18,19], has the highest entropy at ordinary temperature of any form of carbon. Such a perfect crystal is characterized by a certain sequence in the relative position of the closely bonded basal planes of atoms, called ab planes as was established by Bernal[20] Successive ab-planes in the ideal crystal are displaced relative to one another so that every other plane fits exactly over the first. Such a succession may be labeled as ABAB sequence. The bonding energy between planes is only about 2% of that within the plane.[21,22] In α-graphite the planes are so arranged that the carbon atoms of alternate layers are vertically above each other i.e., ABAB sequence. There are other ways of stacking the ab-planes as proposed by Lipsin & stokes.[23] Rhombohedral structure of material appearing to a certain extent in natural graphite in which the ab-planes are ordered on the average as ABCABC and this is called β-graphite. This was originally proposed by Debye & scherrer.[24] However, in β-graphite, the planes are staggered such that the carbon atoms of every third layer are vertically above each other. The β-form is produced by mechanical treatment, such as grinding of the α- form but is converted to the more stable α-form, by heating about 1300°K. ΔH for the transition[25] α-β is 0.14 + 0.04 kcal.g/atom.

The forces that hold the layer planes together are Van der waal's forces, which are only about one-twentieth as strong as the chemical bonds between atoms in the layer planes. Therefore, the tendency for these layers to orient themselves perfectly is not great and does not become effective except at high temperature (~1600°C). On the other hand, high binding energy of the atoms in the layer planes makes them stable when formed, and also permits only slight mobility to correct errors even at high temperatures.

The perfect graphite structure has the carbon atoms in a planar array in sp^2 bonding with no bonding directed out of the layer. Thus each layer is bound to its neighbour by Van der waal's or some other form of weak binding energy. Simplistically, there is no propensity to bond or react in the direction at right angles to the plane. Any reactivity must be seen as deriving from the crystal edges. Any penetration of the crystal surface through the layers, any attempt to bond, must destroy the graphite structure.

Graphite has a very high thermal stability, does not melt and reaches a vapour pressure of 1mm Hg at ~3000°C. Thus, the atom mobility is limited and sintering is not possible. As a consequence of the cleavage on the basal plane and the lack of any directional bonding in the plane, graphite is extremely flexible.

The degree of crystallinity of the graphite[(26)] and its relatives should be recognized to be very dependent on both the temperature of the graphitization and degree of molecular arrangement in the precursor.

The manner of assembly of micro-crystalline and precursor planer molecules can be defined as "Orientation". Such an orientation originates from the layered assembly of planar units which produces strong anisotropy. Planar structure of the graphite is achieved thermally by heat treatment above 2500°C. The structure in Figure 2.2(a) shows graphitizing carbon which possess approximately oriented planar units. The units are combined into graphite layers by heat treatment above 2000°C. In contrast Figure 2.2(b) shows that of so called non-graphitising carbon of which units are distributed randomly. Hence the lattice structure is difficult to grow into graphite.[(27)]

Different types of defects are present in the graphite structure. A brief account of the defects that are generally found are described below.[(28)]

1. Layer stacking defect: Perfect planar network of hexagons of carbon atoms though present as parallel layers, may be displaced from the ideal hexagonal or rhombohedral sequence, thus leading to disordered stacking sequence. This type of carbon is termed as turbostratic.

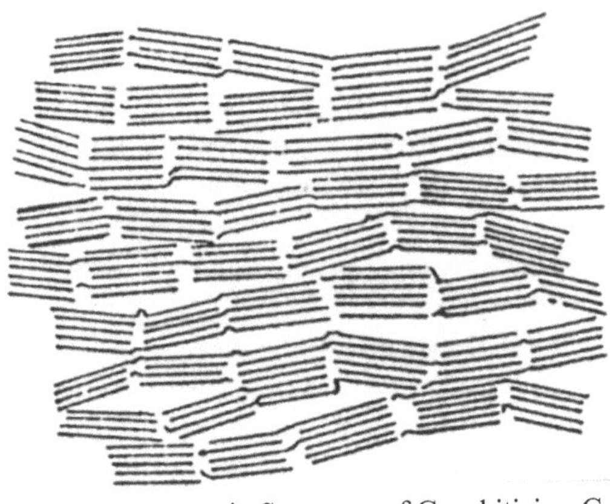

Fig. 2.2(a): Microscopic Structure of Graphitising Carbon

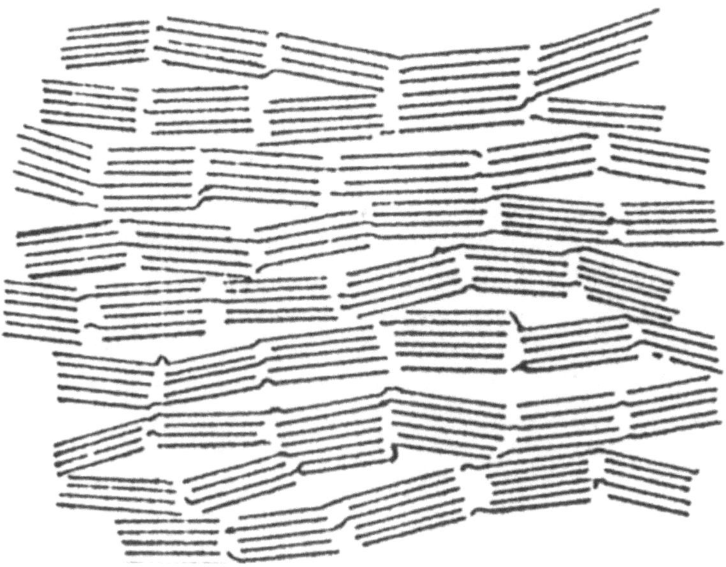

Fig. 2.2(b): Microscopic Structure of Non-graphitising Carbon

2. Carbon network bond defect: A network of finite size is conventionally referred to as carbon macro-molecule which in case of near ideal graphite are aromatic in nature.

The principal type of bond defects are:

(a) ***Edge defect:*** Whenever the C-C bond cannot be formed, for example, one carbon macro-molecule is out of plane with respect to its immediate neighbours but the electronic valence is satisfied by some way or the other, in such cases edge defect is found. This defect arises very often because foreign atoms or groups are bonded at the edge of carbon network and significantly alter the chemical behaviour of carbon. Sometimes the carbons at the edge defects may form weak spin-pairing bond with atoms of neighbouring macro molecules and at high temperatures such weak spin pairing bonds would tend to dissociate and couple with foreign atoms or molecules.

(b) ***Hole or claw defect:*** This type of defect arises when C-C broken bonds lead to holes or rupture in the network of hexagon of carbon. The space limitation in the structure leads to unpaired electrons for one or more of the atoms at the defect site and trap foreign atoms or molecules.

(c) ***Twining defect:*** At the joining of the twining planes, alternative rings with 4 atoms and 8 atoms form. This give rise to deformation in the ring structure. The defect sites in the crystal structure of carbon are responsible for adsorptions of foreign atoms or molecules and serve as the active sites for reaction leaving carbon susceptible to high temperature oxidation in oxidizing atmosphere.

2.2 Gasification of Carbon

A voluminous literature exists regarding the reactivity of carbon in its various forms.[29–48] Many researchers have tried to correlate the rates of carbon gasification to the many physical and chemical properties of the different materials used. Turkdogen *et al.*[34] determined that the difference in the oxidation rate between different types of carbon is mainly due to differences in the effective surface areas. Dekoranyi[47] confirmed this behaviour by showing that the reactivity is linearly related to the micro-porosity. Tien and Turkdogen[37] concluded that because of the strong poisoning effect of CO on the rate of oxidation of carbon, delayed pore diffusion reduced the depth of internal oxidation. Adschiri and Furusawa[43] showed that the rate of CO_2 gasification of coal char in the temperature range from 800°C to 1000°C could be predicted by knowing the initial porosity and initial surface area alone.

In their view, Mishra and Standish[29] concluded that the reactivity of carbon is influenced by several factors including: (i) origin; (ii) pre-treatment; (iii) intrinsic chemical differences in the carbon; (iv) differences in surface area; (v) catalysis by impurities and finally the; (vi) experimental conditions used. Because of all these variables, an universal rate expression on the rate of carbon gasification does not exist.

A piece of carbon when exposed to oxidizing atmosphere, undergoes continuous weight loss. Oxidation of carbon by oxidising agents is popularly known as "gasification" and the rate at which carbon losses weight is generally termed as "reactivity". Gasification reaction is characterized by high activation energy indicative of chemical control of the reaction. However, it has been established that mass transfer also tends to contribute significantly towards the rate of reaction specially at high temperature.[43,48,49,50] As chemical control dominates the gasification reaction, the mechanism of the same has been developed on the basis of adsorption of reactant gases on the surface, reaction on the surface and desorption of the product gas. Langmuir and Hinshelwood[51] proposed the mechanism of gasification on the basis of the above.

It has been found that[52,53] migration of carbon atoms in the solid state is too slow at practically all the temperatures. Thus an oxidation reaction involves:

- initiation step and
- succession steps

In the initiation step, the oxidizing agent gets attached to the part of the carbon network, either on the edge site or on the surface and in that process satisfies the localized unbalanced valences. Following the attachment of oxidizing agent to the carbon network in the "initiation step" of some kind, the solid product will, if the temperature is sufficiently high, rearranges its bond to give a volatile product. This will leave back an active site in the carbon network. Further bond rearrangement may then follow and if oxidizing agent can be attached to the carbon in the same way as before, a succession sequence of oxidation step will be established.

It is apparent from the above discussion that besides other parameters, such as porosity which determines the access of the oxidizing gas inside the sample, surface area which determines the available area for reaction, structural imperfection or defects play a significant role in the kinetics of gasification of carbon. This is because, the extent of defects in the structure of the sample will dictate the number of active sites available. If, somehow, the structural imperfections are removed, then it is evident that there will be reduction in active sites. This, in turn, should improve the oxidation resistance of carbon.

2.2.2 *Thermochemistry of Oxidation of Carbon*

Carbon forms two gaseous oxides[(53a)], CO and CO_2, according to:

$$C_{(gr)} + O_{2(g)} = CO_{2(g)} \qquad \ldots \text{(i)}$$

for which $\Delta G^o_{(i)} = -94{,}200 - 0.20\ T$ calories and

$$2C_{(gr)} + O_{2(g)} = 2CO_{(g)} \qquad \ldots \text{(ii)}$$

for which $\Delta G^o_{(ii)} = -53{,}400 - 41.90\ T$ calories.

Combination of reactions (i) and (ii) gives

$$2CO_{(g)} + O_{2(g)} = 2CO_{2(g)} \qquad \ldots \text{(iii)}$$

for which $\Delta G^o_{(iii)} = 2\Delta G^o_{(i)} - \Delta G^o_{(ii)} = -135{,}000 + 41.50\ T$ calories.

Similarly the equilibrium

$$C_{(gr)} + CO_{2(g)} = 2CO_{(g)} \qquad \ldots \text{(iv)}$$

for which $\Delta G^o_{(iv)} = \Delta G^o_{(ii)} - \Delta G^o_{(i)} = +40{,}800 - 41.70\ T$ calories

For reaction (iv) $\Delta G^o_{(975^o k)} = 0 = -RT \ln Kp = -RT\ln(p^2_{co}/p_{co}{}^2)$, i.e., at 975°K,

$K_p = 1$, and hence $p_{co}^2 = p_{co} = 1$, and $P_{total} = 2$. If the pressure of the system is decreased to 1 atm, then, as Kp is independent of pressure, and hence remains equal to unity, p_{co} becomes greater than p_{co}^2, i.e., the equilibrium of reaction (iv) shifts towards the CO side, as is predicted by Le Chatelier's principle. As $\Delta H^o_{(iv)} = +40{,}800$ calories, reaction (iv) is endothermic, hence, from Le Chatelier's principle, as a decrease in temperature shifts the equilibrium in that direction which involves an evolution of heat, a decrease in temperature shifts the equilibrium towards the $C+CO_2$ side. Thus, if it is required to decrease the pressure of the system from 2 to 1 atm and at the same time maintain $p_{co} = p_{co}^2$, then the temperature of the system must be decreased.

2.3 Mechanism of Gasification

A simplified picture of reaction of a carbon particle with oxygen has been presented in Figure 2.3.

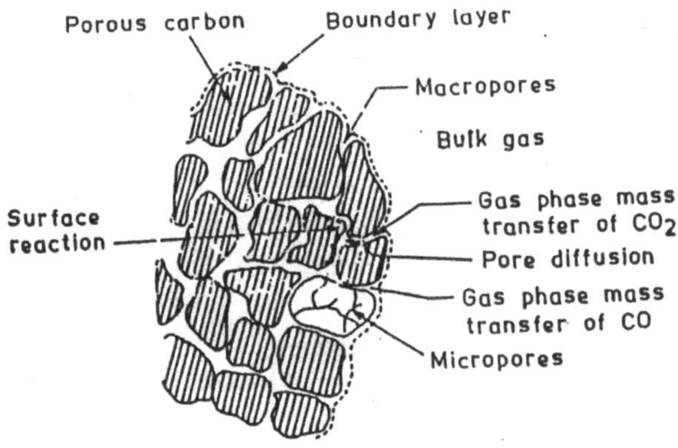

Fig. 2.3: Schematic View of Reaction of Carbon Particle with Oxygen

The kinetic steps[(54)] are:

- Transportation of oxygen from bulk gas to the external surface of carbon particle through boundary layer.
- Diffusion of oxygen into the interior of the pores of the sample i.e., pore diffusion.
- Adsorption of oxygen on the pore surface and external surface.
- Surface chemical reaction in the adsorbed layer.
- Desorption of CO, i.e., the product of reaction.
- Outward diffusion of CO from interior to sample surface through pores.

- Transportation of CO from the sample surface to the bulk gas phase through boundary layer.

Among the kinetic steps mentioned above, steps (c), (d) and (e) are chemical steps as some form of chemical reaction is involved in them and rest of the steps are transport steps. Out of the above, one or more than one steps would be controlling the rate. Many physical variables have profound influence on reaction rate. They may be listed as follows:

- temperature of reaction
- particle size of graphite
- gas composition
- nature of graphite
- impurities associated with graphite.

2.4 Kinetics of Carbon Gasification

The oldest model to treat in gas-solid reaction is the shrinking core model developed by Yagi and Kuni(55) in 1955. This model visualizes the solid reactant as being non-porous and impervious to the gas reactant. The reaction starts from the outer surface and forms a gradually thickening "ash layer" of porous solid product. Gaseous reactants diffuse through this product layer to the sharp reaction interface, which separates the completely reacted and unreacted solid. The gaseous products diffuse back out through the product layer. The progress of reaction is controlled by either external mass transfer to reaction particle or by diffusion in the ash layer, or by the chemical reaction at the interface, or by a mixture of these steps. Early investigation of the model assumed the control by one of these steps.

This explanation was further extended to include the external mass transfer step(56), swelling and shrinking of the solid during reaction(57) and reaction other than first order.(58) If any one of these steps, e.g., external mass transfer to the reaction particle, diffusion through the ash layer and chemical reaction at the interface is rate controlling, a linear dependence of interface radius on time would be expected.

Many authors have worked on solid-gas interaction kinetics. In working with reduction of iron oxide with carbon monoxide, Mckewan and Smith-(59) assumed reaction as topochemical and chemically controlled. Mathur, Rao and Ghosh(60) assumed a single interface topochemical reaction pattern with gas diffusion through porous product layer as the rate controlling mechanism.

There are many shortcomings in the topochemical shrinking core model. Frequently, solid reactants are porous and therefore not impervious to gaseous reactants. Many investigators[61,62] have observed a diffused reaction zone separating completely reacted and completely unreacted solid instead of sharp interface in the partly reacted porous solids. These shortcomings of the shrinking core model have led to the development of diffuse interface model by Ausman and Watson.[63,64] This model visualizes each solid particle as an agglomerate of uniform sized spherical grains and each grain is reacting in the form of shrinking core as shown in Figure 2.4. It was shown that smaller the grain or pores, with constant porosity, the faster the reaction rate. Also, the greater the porosity, the faster is the reaction rate. A large number of authors[65–70], since then, have published, diffuse "interface model" and described their application in reduction of oxides.

Many researchers have tried to correlate the rate of gasification to many physical and chemical properties of the material used. Turkdogan *et al.*[71] determined that the difference in oxidation rate between different types of material is due to difference in the effective surface areas Dekoranyi[72] confirmed this behaviour by showing that the reactivity is linearly related to the microporosity. Tien and Turkdogan[73] concluded that because of the strong poisoning effect of CO on the rate of oxidation of carbon, delayed pore diffusion reduced the depth of internal oxidation. Adschiri and Furusawa[74] showed that the rate of CO_2 gasification of coal char in the temperature range from 1073 to 1273°K, could be predicted by knowing the initial porosity and initial surface area alone.

A few investigators proposed previously the models for non spherical particles such as the cylindrical and plate models based on the unreacted core model.[75–76] The thickness of the product layer at a certain conversion is assumed to be same regardless of the location of the layer in these models. It means that the reaction interface moves at the same velocity regardless of the location of the interface at a certain conversion. Sohn *et al.* have proposed the structural model[77] in which the shape of the particle and the subgrain are classified into sphere, cylinder and disc but there is no definition about geometry, so it is difficult to utilize this model. However, there is no unanimity about these.

2.4.1 *Mass and Heat Transfer Limitation on Rate*

Kinetic steps of gasification reaction include transportation of the reactant and product gases to and from the reaction sites. Some workers[78–81]under specific experimental conditions noticed significant effect of mass transfer step on gasification kinetics. Walker *et al.*[81] discussed the role of transfer of reactant and product on gas carbon reaction. According to Rossberg[79] three distinct regions may be identified. Role of diffusion on each region may be described as follows:

- At low temperature the reaction is generally chemically controlled, mass transfer does not play any significant role in determining the kinetics of the reaction. At this stage, the whole porous body experiences attack from reacting gas that prevails in the bulk.
- At high temperature, the resistance offered by mass transfer steps for reactant and product gases are no more negligible. With increases of temperature, the rate of the chemical step of a reaction increases rapidly but the relative increase in the mass transfer steps are much less. Thus the mass transfer and chemical reaction rates become comparable and the gas composition in the porous solid varies considerably from that of the bulk composition. Under this situation, each unit volume does not register the same rate of reaction as in the case of earlier one.
- At still higher temperature the chemical step becomes much faster compared to mass transfer and offers no resistance to the progress of reaction. The surface concentration of reactant gas becomes a fraction of bulk concentration and penetration takes place to a limited extent only.

Thus the reaction virtually becomes confined to the surface. So, it is apparent from the above, that mass transfer is expected to play an important role in the gasification reaction at high temperature as well as intermediate temperatures. As the particle size increases, the diffusional path length also increases and chance of mass transfer controlling the rate becomes more. At high temperature, an exclusive chemical control reaction can be achieved only by the use of small particles. On the other hand, at lower temperature, due to slow chemical reaction rate, reactions have exclusive chemical control even for somewhat larger particles.

Many investigations[82–87] have been carried out to find out the influence of particle size on the gasification rate of carbon. Rao[83] found that rate increases when weight of sample is increased. Standish *et al.*[84] have found that time for complete conversion increases with increase of particle size. But Ergun [85] found that variation of particle size from 0.2 to 1 mm did not affect the gasification rate. Van Heck *et al.* [88] also observed no effect of particle size when size was varied from 0.1 to 1.4 mm. Literature reports that effect of particle size on gasification of coal char is complex [89] and has been found to range from nil[90] to 35 folds.[86]

Turkdogan *et al.*[86] and Tien and Turkdogan[87] done extensive work on effect of particle size on gasification rate for different types of carbon. Turkdogan *et al.*[82] also tried to distinguish the regions of "complete internal burning", "partial internal burning" etc. as a function of particle size. They found that at lower temperature the zone of chemical control is extended upto higher particle size than at higher temperature.

Carbon gasification reaction is endothermic in nature. So, if the rate of heat transfer through the material is not rapid enough, the central part of the sample may be at lower temperature than that of surface. The temperature difference between the surface and the center will be high with large particle size. Again at higher temperature as the reaction rate is greatly enhanced, it is expected that temperature difference will be significant. Thus for larger particles and at high temperature there may exist a significant temperature gradient from surface to center depending on thermal conductivity of bed. Porous and powdery mass are more prone to temperature gradient due to poor conductivity.

2.4.2 *Surface Area Effect*

In the structure of graphite porosity always exists and it varies in amount and size and can be classified into (a) micropores (10Å), (b) Semi micropores (10–300 Å) and (c) macropores (>300 Å). Microporosity in carbon arises due to imperfections in structure. The most common technique of surface area measurement is by BET using adsorption of nitrogen gas at liquid nitrogen temperature.

Rate of gasification varies with carbon burn off. This aspect has received attention from several investigators.[91–95] But the nature of variation of rate reported by various workers does not follow the same trend. However, in a large number of cases it has been found to increase to a maximum at around 40 to 60 percent burn off and then decreases again.

In order to explain this, many workers determined changes in pore surface area with burn off. It is the general view of the workers[91,95] that at the initial stages of reaction, the micropores once available become accessible to the gas and this gives rise to increase in surface area. But with progress of gasification, the micropores collapse to bigger size of pores resulting in decrease of surface area.

In order to tackle the problem of dependence of gasification rate on physical parameters like surface area, porosity, etc. some workers have tried to formulate empirical equation giving relationship of $\{-(dW_c/dt/Wc)\}$ with fractional conversion.[92,95,96] However, there is no confirmation about these.

Turkdogan *et al.*[91] tried to correlate reaction rate with surface area for different types of carbon. It may be noted that the rate is proportional to the specific surface area for pyrolytic graphite, electrode graphite and coke. However, for coconut char, actual rate is much lower than extrapolated rate. He ascribed this behaviour of char to predominance of micropores in it. According to him, gases do not diffuse into the micropores easily. So, micropores basically remain inactive during gasification reaction.

2.4.3 *Effect of Catalysis on Gasification Reaction*

Gasification reaction of carbon is found to be influenced quite significantly by foreign elements or compounds. Many investigations have been carried out on catalytic gasification of carbon. Investigators have found that transition metal, alkaline earth metals etc. greatly enhance the reaction rate. A comparative study on the relative activity of various types of metal catalysts was carried out by Letort *et al.*[97] in the combustion of graphite in air at 773°K in presence of 0.12% catalyst. They presented the following series in increasing order of catalytic activity.

$$Au > Na > Ca > Ag > Cu > Pb > Ni > Mn > Co > Fe$$

But one thing must be borne in mind that the relative activities are by no means universal but depend strongly on the particular experimental conditions like size and porosity of catalyst particles (and thus their surface area),chemical state of the catalyst, intimacy of contact between catalyst and carbon surface etc. On the other hand sulphur bearing compounds have proven to be inhibitors of gasification reaction. Walker *et al.*[98] compiled various aspects of catalytic reaction of carbon. The theories advanced to account for the experimental findings of catalytic gasification fall broadly into two categories. In brief they can be described as follows:

2.4.3.1 *Oxygen Transfer Mechanism*

This mechanism was first proposed by Neumann *et al.* [99] And then by Milner *et al.* [100] to explain catalytic gasification reaction. The scheme proposed by Milner *et al.* may be presented as:

$$RO + CO_2 \rightarrow RO.CO_2 \quad \ldots (1)$$

$$RO.CO_2 + C \rightarrow RO + 2CO \quad \ldots (2)$$

Where RO represents metallic oxides.

Vestola and Walker[101] expressed views similar to the above in principle, and differed only in the chemical entity of the intermediate compounds. Catalysts on the carbon surface are regarded as oxygen carriers by an oxidation reduction cycle. Elements or compounds, which do not undergo such cycle under the experimental conditions, cannot exhibit any catalytic effect.

2.4.3.2 *Electron Transfer Mechanism*

The electronic theory of catalysis of gas-carbon reaction envisages a transfer of electrons between two solid phases. Long and Sykes[102,103] proposed that oxygen atom gets adsorbed at the active sites of the carbon network and changes the distribution of π-bonds. This oxygen–carbon bond is very weak and readily decomposes to carbon monoxide. Turkdogan and Vinters[104] studied the rate of gasification of iron impregnated graphite granules (<0.01 to 2.1% Fe) in CO-CO_2 mixture at temperature range of 973–1273°K and at pressure of 0.03 to 1 atmosphere. They also carried out gasification of iron impregnated coconut char and coke. Other catalyst elements like Ag, Cu, Cr, Zn, Ni, Co were used for study of gasification of graphite granules in air. They observed that even with less than 0.01% iron, the rate of reaction of graphite in CO-CO_2 gas mixture at 973–1273°K increased by a factor of 2000. Presence of 2% impregnated iron enhanced the gasification rate at 1073°K by a factor of around one million. One interesting feature of their study is that they observed marked decrease of catalytic capability of transitional metals beyond 20 to 50 pct. conversion. However, hydrogen treatment of partially reduced graphite revived the catalytic effect. This finding is in accordance with what has been reported by walker *et al.*[98] Work by Mehrotra and Sinha[105] shows that there exists a marked difference in degree of catalysis of iron on the gasification reaction. They used 0–15% iron for oxidation of carbon black in carbon dioxide atmosphere in the temperature range of 1073–1273°K. They observed only 2.5–3.0 times rate enhancement with 15% iron which is insignificant as compared to what was reported by Turkdogan *et al.* The reason for such a huge difference of extent of catalysis by iron can be explained in the following manner. Turkdogan *et al.* used impregnated iron in the carbon sample. With impregnation, iron atoms got incorporated into the structure of carbon and were finely disseminated at molecular level resulting into a very large Fe-C contact area. On the other hand, Mehrotra and Sinha employed a mechanical mixture of iron carbon. Thus iron – carbon contact area was much less than that of impregnation. So for catalytic effect, the more important factor is the contact area rather than the amount. To support this point, low rank coal may contain [106] relatively high amount of alkaline oxide in the structure. Presence of catalyst cat ion Ca^{+2}, Na^{+}, K^{+} etc. (i.e. exchangeable cat ion) in high proportion in lignite char is probably the cause of high reactivity. Lignite coal contains carboxyl functional groups in their structure and exchangeable cat ions are thought to be present with these carboxyl groups. So the cat ions are intimately mixed and uniformly distributed in the carbonaceous matrix and thus catalyses gasification reaction more effectively. Treatment of char with acids (HCl, HF) removes the exchangeable cat ions and thus demineralized chars are less reactive than original chars. Concentration of carboxyl functional groups decrease with increase of coal rank. Therefore, inherent catalytic effect may not be significant for higher rank coals.

Otto *et al.* [107] found that 0.001–1 atomic percent of Pb can increase the gasification rate of graphite by around 5 order of magnitude, primarily because of massive increase of density of reaction sites. Based on absolute reaction rate theory, they found that site density was consistent with geometry of graphite crystal and postulated that every carbon atom in the structure can be transformed into a reaction site.

Many investigations[108–114] on catalytic effect of alkali metal compounds have been reported recently. From their kinetic data on gasification of carbon in carbon dioxide with 5 weight pct. of K_2CO_3, Wigmans *et al.* [110] found the electron transfer mechanism to be valid to explain the catalytic effect of K_2CO_3. According to Cerfontain *et al.*,[111] the difference of extent of catalytic effect of Na, K is due to the difference in amount of chemiabsorbed oxygen on the carbon surface. Freund [108,109] has found that even though Ca, K catalyzed the gasification reaction to a significant extent, they do not show any appreciable effect on decomposition of carbon-oxygen complex and exhibit the same activation energy for both catalytic and uncatalytic gasification reaction. Thus he has concluded that enhancement of gasification reaction of carbon is due to increase of active carbon sites in presence of catalysts. This finding is in accordance with the report of Otto *et al.*[107]

Alam and DebRoy[113] have made a very comprehensive study on gasification of coke in carbon dioxide using KCN, KOCN and K_2CO_3 as catalysts. They have observed that the potassium bearing compounds enhance the rate quite significantly. They have also noticed that extent of enhancement do not depend upon the anionic constituents i.e. KCN, KOCN and K_2CO_3 have same degree of catalyzing effect. They have explained their catalysis data on the basis of oxidation-reduction cycle mechanism i.e. red ox mechanism of precursors. With the help of micrographs of partially reacted coke sample, they have found potassium to be evenly distributed in the carbon matrix. They have attributed the extensive covering of surface by potassium to the transport of the same to the interior of small pores via gaseous medium either in elemental form or as volatile compound. Alam *et al.*(113) have also measured the vaporization loss of potassium as a function of carbon burn off. Molar K/C ratio has been found to decrease at the initial stage (up to around 40%) of coke conversion and after that the ratio increases. The increase of molar K/C ratio at the advanced stage of burn off has been explained by faster rate of gasification over the rate of loss of potassium. They have tried to normalize the gasification rate per unit amount of potassium as a function of coke conversion. To eliminate the effect of structural parameter, they have plotted gasification rate of coke per unit surface area against coke conversion. It may be noted from the figure that behavior of specific rate for catalyzed samples are somewhat different from uncatalyzed sample. Rate is found to decrease first at the initial stage, then remain constant up to around 60% burn off, and finally increases again.

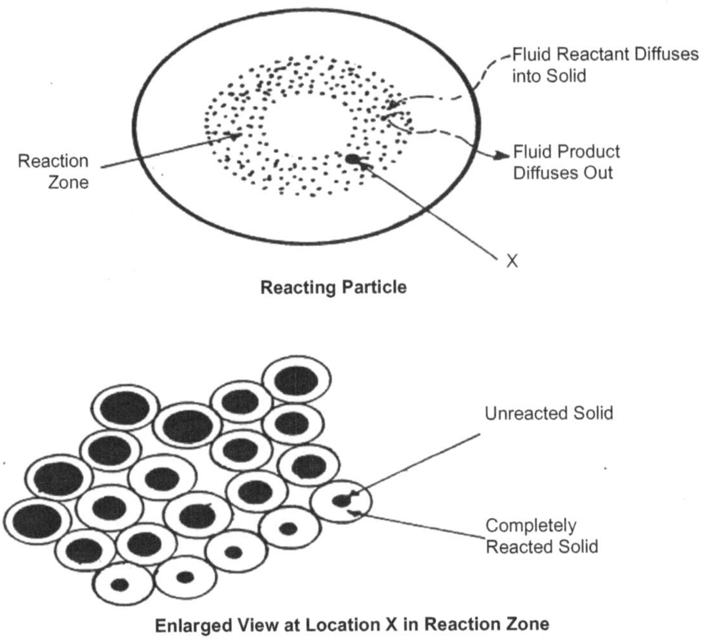

Fig. 2.4: Schematic Diagram of Diffused Interface Model

CHAPTER 3

EXPERIMENTAL DETAILS

3.1 Experimental Details

Experimental investigation performed initially for characterisation of graphite samples and then study the oxidation behaviour in air and oxygen atmosphere by studying the reaction kinetics. The temperatures of oxidation are so chosen as it has been observed by DTA/TGA analysis that maximum oxidation taking place at 650–700°C. The detail tests carried out are outlined below:

1. ***Proximate analysis:*** Determination of moisture, volatile matter, ash, fixed carbon.
2. ***Particle size:*** Determine the size distribution of grains in the tested samples.
3. Specific gravity of graphite.
4. ***Specific surface area:*** Determine the fineness of material to understand the activity of material.
5. ***Chemical analysis of ash:*** Determination of elemental oxides, sulphur, etc. present in the ash.
6. ***Thermal analysis:*** Differential thermal analysis and thermo-gravimetric analysis to study the thermal behaviour of graphites.
7. ***Scanning electron microscopy:*** Determine the impurity, cleavage in the grain, pore and crack distribution and also avg. diameter of particle.
8. ***X-ray diffraction study:*** Calculation of '*d*' spacing and also for degree of graphitisation.
9. ***Petrography study:*** To study the nature of grain.
10. ***Fusion temperature of ash:*** Determine the softening point of ash.
11. ***Oxidation rate study of graphite:*** Oxidation of graphite in air/oxygen.

3.2 Graphite Sampling

For assessing the property representative samples should have to be collected. Much depends on the correct collection of samples. The word 'sample' means a small portion of a mass, which accurately represents the mean composition of the whole mass. Sampling is thus, a process of taking a small portion of an article such that the consistency of that portion should be representative of the whole.

The important conditions affecting the selection of a true average sample from a large bulk of material are:

- Thorough mixing of the bulk.
- The fineness of the division of the sizes of the constituents forming the bulk.
- The weight of the sample taken is directly dependent upon the size or degree of fineness of the ore particle.
- The ratio of the weight of the first sample to the total weight of the bulk. It is usual to cut out for first sample from one-fifth to one twentieth of the bulk.

The process generally followed here is manual coning and quartering. The ore is mixed well by turning it over several times with shovel and then the ore-heap is piled to form a cone. The even distribution of the material is absolutely necessary to obtain a proper sample. In building up the cone, the fine material always form the apex, while the coarser particles tend to fall towards the base. On subsequent leveling and quartering the cone, this local richness may be partly corrected, according to whether less or more of the fine ore forming the apex is shoveled into one of the sections, and not evenly distributed among all the four sections.

The cone then divided into four equal parts by pressing a cross formed from iron plates riveted together. Two opposite quarters then cut with shovel and removed and other two are then formed into a fresh cone; which is again quartered as before. The process is repeated two or three times until the sample becomes a kilogram in weight. Samples investigated were the available samples of the three sources supplied by the respective manufacturers. Physico-chemical behavior may vary to an extent depending on the fixed carbon content of the graphite minerals.

3.3 Proximate Analysis

The proximate analysis is essentially an examination of the suitability of graphite for further commercial use. The proximate analysis involves the determination of the overall composition of graphite and in reality, the determination of moisture content, volatile matter, ash content and fixed carbon content by differentiation. Proximate analysis were carried out according to ASTM standard D3172.

3.3.1 *Sample Preparation*

The collected sample were first spread on tray for drying. The dried and powdered samples were then passed through 250 μm (72 mesh) sieve. The powdered samples were stored immediately in airtight container to avoid water absorption. Before each measurement, the samples were again dried in order to ensure absolute dryness.

3.3.2 *Moisture Content*

The moisture content of graphite was determined as per ASTM standard D 3173 by measuring the percentage weight loss of the sample. The sample (1 gm) was maintained under controlled conditions (107 $\pm$ 3°C) in an inert atmosphere for 1 hour. Moisture content was determined by the following formula:

$$\text{Moisture content, } \lambda = (W_1 + W_2) \times 100/W$$

Where, W = weight of the sample before air-drying.
W_1 = weight loss in air
W_2 = weight loss in oven

3.3.3 *Volatile Matter*

The volatile matter content was determined by measuring the loss in weight percent. About 1 gm of moisture free graphite sample was put into the reactor and air was expelled from the tube by passing argon gas for about ten minutes. The air free reactor tube containing the sample under the flow of argon gas was put inside the tubular furnace, which has already been heated to a temperature of 950°C. After 20 minutes, the reactor tube was removed from the furnace and was allowed to cool under argon flow. The difference between the weight of the sample before and after the experiment was considered as the weight loss due to emission of volatile material.

$$\text{Volatile matter, \%} = W - W_1 \times 100/W,$$

Where, W = weight of the sample before experiment.
W_1 = weight of the sample after experiment

3.3.4 *Ash Content*

The ash content is defined as the incombustible material which remains when a graphite sample is burnt. The amount of ash was determined by burning (1–2 gm) of a graphite sample in an adequately ventilated muffle furnace and was gradually heated to redness at such a rate as to avoid mechanical loss and rapid expulsion of volatile matter. The ignition was continued until a constant weight is reached at a temperature of 950°C. The crucible was then allowed to cool in a desiccator and the weight was taken as soon as it became cold. Ash content was determined by using the formula:

Ash content (%): (weight of ash) × 100/weight of moisture free graphite sample taken.

3.3.5 *Fixed Carbon*

The fixed carbon value was determined as per ASTM D3172 by subtracting from 100, the resultant summation of moisture percent, volatile matter percent and ash percent, i.e., Fixed carbon = 100 – (ash% + volatile matter% + moisture%).

3.4 Particle Size Measurement

Particle size of the finer fractions of samples was measured in the instrument working with the principle of Laser diffraction. The apparatus was supplied by M/s Galai products Ltd., Israel, model number CIS-I. Here, sample was kept in a cuvette of quartz glass of size (1 × 1 × 3) and capacity was 3 ml. Solution used as suspension was tetrachloroethylene. Sample size of (0–150 micron) can be measured in this instrument. It measures the shadow of the particle with time transient method. One magnetic stirrer was provided for uniform movement of particles and results were obtained in tabular/graphical mode.

3.5 Specific Gravity Measurement of Samples

Specific gravity of samples were measured by the standard specific gravity bottle following ASTM C-127.

3.6 Surface Area Measurement

An important method for measurement of surface area of fine powders is the standard low temperature nitrogen adsorption method according to Brunauer, Emmet and Teller.

3.7 Chemical Analysis by XRF

Quantitative estimates of the amounts of the elements present in a sample was measured by X-ray fluorescence method.

3.8 Differential Thermal and Gravimetric Analysis

The technique of differential thermal analysis is an important tool to study the structural changes occurring both in solid and liquid material under heat treatment.[116] These changes may be due to dehydration, transition from one crystalline variety to another, destruction of crystalline lattice, oxidation, decomposition etc. The principal of DTA consist of measuring the heat changes associated with the physical or chemical changes occurring when any substance is gradually heated. This technique has attained considerable importance in determining the oxidation behavior of carbonaceous materials.

3.8.1 *Working Principle*

When a sample and a reference substance are heated or cooled at a constant rate under identical environment, their temperature difference is measured as a function of time or temperature. The temperature of the reference substance, which is thermally inactive rises uniformly when heated. DTA is the process of accurately measuring the difference in the temperature between a thermocouple embedded to a sample and a thermocouple in a standard inert material, such as aluminum oxide while both are being heated at a uniform rate. The DTA method involves the simultaneous recording of temperature (T) which exists in the furnace and of the temperature difference (ΔT) which appears between the sample and thermally inert material. Their differences of temperature arise due to the phase transitions or chemical reactions in the sample involving the evolution of heat (exothermic reaction) or absorption of heat (endothermic reaction). The exothermic reaction and endothermic reactions are generally shown in the DTA traces as positive and negative deviations respectively from a base line. So, DTA gives a continuous thermal record of reactions in a sample. The areas under the peaks are proportional to the amounts of active material present.

3.8.2 *Apparatus*

The apparatus of the differential thermal analysis consists of a thin walled refractory specimen holder made of sintered alumina with two adjacent cubical compartments of exactly the same size 1 cm in length. One for the reference (inert) material and the other for the test material, the compartments being separated by a 1 mm wall. The specimen holder is placed in the cavity of the heating block, which is made of fine grained refractory cement. The block is operated in the center of the cylindrical refractory tube of an electrical furnace, which supplies an uniform heating rate. The furnace (9″ × 6″ × 9″) is packed with calcined china clay. The input of current into the furnace is secured through the secondary of a variac transformer, which controls the current. Fine chromel-alumel wires (28 gauge) are used for thermocouples. A cold junction is used for thermocouple leads and the e.m.f. is recorded almost continuously, while the temperature of the inert material is measured at 3 minutes interval shown in Figure 3.1(a). It is essential to use perfectly dry materials as otherwise errors will be introduced in the analysis. Approximately 0.1 gm anhydrous alumina is used in the reference cup and the sample weights varies over a range of 0.05 gm to 0.125 gm depending on their packed density. A heating rate of 10°C per minute (average) of the furnaces is conveniently kept, and this gives satisfactory results in most cases. A block diagram of DTA is shown in Figure 3.1(b).

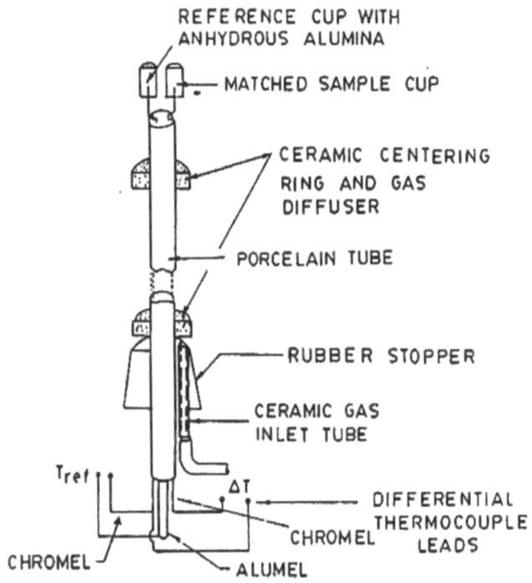

Fig. 3.1(a): DTA Thermocouple Assembly

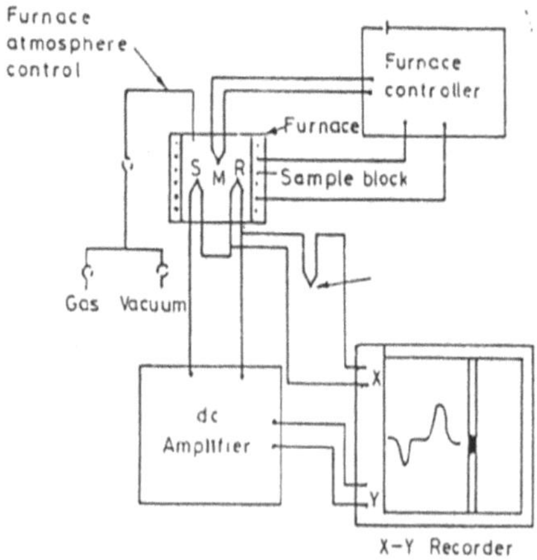

Fig. 3.1(b): Block Diagram of DTA

3.8.3 *Thermo-Gravimetric Analysis*

The thermo-gravimetric analysis (TGA) involves the determination of weight loss for a sample as a function of time or temperature, i.e. $f(t$ and $T)$, while a sample is heated or cooled at a constant rate. This technique is effective for quantitative analysis of thermal reactions that are accompanied by mass changes due to release of volatile matter, evaporation, decomposition, gas absorption and dehydration.

The rate of decomposition varies with the heating rate. The maximum decomposition rate occurs at a higher temperature when a graphite sample is heated up at a faster rate. The maximum rate of weight loss appears at a higher temperature for graphite i.e., at about 650°C for all types of graphite.

Powder graphite samples of sizes (–30 to 60) BSS and (–60 + 150) BSS were used for thermo-gravimetric analysis. About 40 mg. of moisture free sample was taken initially in a platinum crucible. Thermo-gravimetric investigations were carried out by determining the rate of loss in weight as a function of temperature at a constant heating rate, i.e., 10°K/min.

3.9 Scanning Electron Microscopy

In electron microscopy, superior resolving power of a beam of electron, is used to investigate the shapes and sizes of crystal at a much finer level than is possible with the light microscope. One of the most useful capabilities of the electron microscope is the ability to form electron diffraction pattern from very small region of the sample. Combination of electron diffraction evidence from a substance with morphological evidence from electron micrograph of the same region can be invaluable in relating growth habits to crystal structure features or in studying epitaxial or topotactical relationships between crystalline phases. The microscope used for the study was JEOL Jsm-840A.

3.10 X-ray Diffraction

X-ray diffractometry permits the detection, identification, and determination of the concentration of crystalline phases, such as minerals. In XRD, a beam of X-rays from a high powered tube is directed at a smooth surface, produced by a former – scraper, and a detector is used to measure the X-rays diffracted through angles characteristics of the crystalline phases present. Bruker-AXS Model No. D8 advance X-ray diffractometer was taken and sample was put into holder by making paste with pressure. Sample is first made into fine powder and passed through 200 B.S. sieve.

The material was taken as a thin layer on a plastic substrate in an area of 2.0 cm^2. X-ray powder diffraction was carried out using a two-circle diffractometer with Cu K_α radiation at 25 mA/35 KV.

3.10.1 *Graphitization and Its Measurement by XRD*

X-ray diffraction has been extensively used to characterize the graphite samples. The following changes in diffraction pattern occur with the progress of graphitisation.

- The narrowing of the (001) diffraction lines and 2θ shifts.
- The transition of two dimensional lattice type (hk) reflection into crystalline (kk1) reflection.
- Increase in preferred orientation.

Different parameters are used to express the degree of alignment of adjacent layers, e.g. degree of graphitisation (g), probability factor (p), etc. Maire and Merings '*g*' factor is most widely used parameter to characterise the degree of graphitisation and is given by: $a_3 = 3.354 + 0.086\ (1\text{-g})$.

Where a_3 = spacing between adjacent layers. So, determining a_3 by XRD, '*g*' the degree of graphitisation can be determined.

3.11 Petrography Study

At first the sample of graphite was broken into fine powder. Then with jelly like sticky canada balsam the sample was cooked. The canada balsam is colourless, isotropic in character and have a known refractive index of 1.35. We got a mounded paste like specimen. After constant observation and cooling, the mounded slide is covered with transparent coverslip. These slides were kept in cold condition.

3.11.1 *Optical Microscopy and Observation*

The samples were placed on the stage of a microscope model no. Leitz Orthoplan-Pol, Germany, equipped with Leitz vario Orthomat two, 35 mm wild MPS 15 semi photomate, photomicrograpic camera and photograph was taken by using transmitted light. The light source of microscope was a 12 V, 100 W, halogen bulb. Kodak colour films HR 400 were used. An automatic exposure time was programmed depending on the intensity of the transmitted light.

3.12 Determination of Fusion Point of Ash and its Phase Analysis

Three graphite samples were burnt in air in a muffle furnace at 1000°C for 3 hrs by evenly and thinly spreading the samples on a flat refractory surface and keeping it inside the furnace. After cooling the furnace, the samples were taken out and ash samples were collected. The three samples were then tested in a Leitz heating microscope at rising temperature and change of sample morphology noticed. The temperature at which the first deformation takes place was noted and that is the fusion temperature of the sample and the temperature increased further and the temperature at which all the samples becomes hemispherical in shape noted and that is the melting temperature. All the samples were then studied under x-ray machine of M/s Bruker-AXS, model D8, Germany with copper as target under K_α radiation and X-ray diffraction pattern were obtained for three samples for phase identification of the ash samples.

3.13 Oxidation Rates Study

3.13.1 *Apparatus for Measurement of Oxidation Rate of Graphite in Air and Oxygen*

Oxidation rate of graphite were measured in a conventional thermo-gravimetry setup model NETZSCH, STA-409, Germany. The set-up consisted of the following:

- Commercially available purified gas cylinders, i.e., IOLAR Gr1 (> 99% purity).
- The thermo-gravimetry apparatus with furnace, reaction chamber, temperature measurement and control unit, facility for continuous measurement of weight loss.

3.13.2 *Gas System*

The purpose of the purified gas cylinders is to use impurity free gas. Argon was used for flushing the reaction chamber before and after the experiments to ensure an inert atmosphere in the chamber. Oxygen and air used for oxidation of graphite.

The air & oxygen used for oxidation was passed through flowmeter where the flow rate of gas were registered. Gases emerging out from flow meter finally entered into the reaction chamber.

3.13.3 *Thermo Gravimetry Set-up*

It consists of a vertical furnace, a mullite tube as reaction chamber, the housing of sample holder, gas inlet and outlet pipes, thermocouples for controlling and measurement of temperature, sample hanging assembly and a single pan semi-micro balance for continuous recording of instantaneous weight. Schematic diagram of the apparatus is shown in Figure 3.2.

A Kanthal wound vertical furnace of 53 cm length was used in the study. A mullite tube of 62 cm length and 5 cm internal diameter was employed as reaction chamber. The mullite tube reaction chamber was fitted with two brass flanges (one at the top and another at the bottom) to provide air tight fitting covers. Both the flanges were having copper coil brazed on them for water cooling.

Fig. 3.2: View of TGA Apparatus

Gas outlet tube is made of stainless steel with a hood over the sample holder. A glass cover was provided on the gas outlet pipe. Its purpose was to prevent any forced vibration to the hanging assembly due to external disturbances such as thermal convection. The single pan balance was placed on a wooden platform with a central hole. An inconel rod having a cylindrical weight at its lower end was connected to the balance. The sample bucket holder was suspended to the constant temperature zone of the reactor tube with the help of the inconel rod. Open end of the glass tube placed over the stainless steel gas outlet tube was sealed with a copper disc having a central hole in it—to allow the inconel rod connecting the balance to pass through. A copper tube around 8 cm long was attached around the hole of copper disc to reduce the effective cross sectional area of the opening. This arrangement was found to cut down the back diffusion of air.

The furnace had about 6 cm long uniform temperature zone, where the sample could be conveniently located. One of the two thermocouples was used for temperature control and the other one for temperature measurement of reaction chamber. The thermocouple was placed inside the reaction chamber rather than near the outer surface for better temperature control. The measurement thermocouple was connected to either a Leeds and Northrup model 8686 millivolt potentiometer or to a Digital milli voltmeter to measure the actual temperature of reaction. Both the thermocouples are calibrated against standard *Pt/Pt-Rh* thermocouple.

It was necessary that the gas temperature and the sample temperature should not differ much for an isothermal study. This necessitated that the flowing gas should have opportunity to get preheated to the reaction temperature before coming in contact with the graphite sample. Heat transfer calculations shows that a separate preheating furnace was not necessary.[(64)] However, in order to ensure attainment of temperature, the gas was introduced through a specially designed inlet tube. A copper tube of 2.5 mm internal diameter was closed at the top and three holes were made at the side wall, one beneath the other. The copper tube was positioned through one of the bottom holes such that the holes of the copper tube were facing the furnace tube wall. This served as gas inlet tube. The incoming gas was forced through the holes of the tube and was thus made to impinge on the heated wall of the furnace. This would result in swirling motion of the gas. Due to these there would be better heat transfer between furnace wall and gas resulting in improved preheating efficiency.

3.13.4 *Material Preparation and Experimental Procedure*

The raw material used for the experimental studies, i.e., three graphite samples were collected from the suppliers. Graphite powders obtained was screened into the fractions (–30 + 60) and (–60 + 150) BSS using screens as these sizes are generally used for commercial purposes. Intermittently the powders were heated in oven at 120°C for 3 hrs to remove moisture to avoid sticking. The sized fractions were then treated with magnet manually to remove any exogeneous iron contamination and finally stored in desiccators for further use.

3.13.5 *Experiments Procedure*

The experimental schedule consisted of the following:

- Oxidation rate study of Palamau graphite (PG1 and PG2) in pure air at 735°C, 785°C and 830°C.
- Oxidation rate study of Bolangir graphite (BG1 and BG2) in pure air at 735°C, 785°C and 830°C.
- Oxidation rate study of Tamilnadu graphite (TG1 and TG2) in pure air at 735°C, 785°C and 830°C.
- Oxidation rate study of Palamau graphite (PG1 and PG2) in pure oxygen at 735°C, 785°C and 830°C.
- Oxidation rate study of Bolangir graphite (BG1 and BG2) in pure oxygen at 735°C, 785°C and 830°C.
- Oxidation rate study of Tamilnadu graphite (TG1 and TG2) in pure oxygen at 735°C, 785°C and 830°C.

3.13.6 *Oxidation Rate Study of Graphite in Oxygen & Air*

Graphite powder sample of sizes (–30 + 60) and (–60 + 150) BSS were taken in a properly cleaned cylindrical 0.5 cm internal diameter, 1 cm. length alumina bucket. The depth of the bucket with sample were kept constant for all the tests to nullify the effect of bed depth. Before putting the graphite powder, the bucket was cleaned with acetone and was accurately weighed. Preweighed bucket was then filled with graphite powder. Outer surface of bucket was then cleaned and bucket was weighed accurately to find out the amount of sample taken.

The reaction chamber of the furnace was flushed with the purified argon for at least 5 min. to remove air and thus ensuring neutral atmosphere. After flushing, the top inlet of the furnace was opened and the sample holder alongwith sample was hanged inside the chamber by hooking it with the inconel rod coming from the balance. The top of the furnace was then closed and the single pan balance was released. After inserting the sample, argon gas was passed for another five minutes. Immediately after five minutes of argon flow, weight of the hanging assembly along with the sample was registered. The weight was taken as zero weight. Then the furnace was heated up and temperature raised to testing temperature in argon atmosphere. After attaining the desired temperature argon flow was then cut off and flow of oxygen and air as per the requirement at required flow rate set in. Instantaneous weight of the sample was recorded continuously. The experiments were continued with graphite until constant weight was reached. Then the oxygen/air gas flow was cut off and argon flow was switched on simultaneously.

CHAPTER 4

RESULTS AND DISCUSSION

4.1 Introduction

Graphite occurs in nature along with other minerals and impurities with very low fixed carbon. With adoption of crushing, grinding and beneficiation, the percentage of fixed carbon increases and depending upon the process adopted very high purity graphite can be produced. One of the main user of graphite is the refractory industry to manufacture carbon containing refractories, and it consumes major portion of the beneficiated product. In India, mainly there sources of graphite are used by the refractory industry and they are Palamau, Bolangir and Tamilnadu grade and for our investigation these three sources were used. The three sources of graphite of fixed carbon content and grain sizes mentioned here were chosen as these quality are used for refractory manufacture by Indian manufacturers. The investigation carried out in the study can be divided into two categories. One is the physico-chemical characterisation of material and other is oxidation behaviour

4.2 Proximate Analysis

Proximate Analysis reveal that all the threc graphites are having high carbon content. Details of the analysis of the graphite samples are given in Table 4.1.

Table 4.1: Proximate Analysis of Three Graphite Samples

Sample No.	*Sample Size (BSS)*	*Moisture (%)*	*Volatile Matter (%)*	*Ash (%)*	*Fixed Carbon (%)*
Palamau	–60 + 150	0.23	2.46	18.21	79.1
	–30 + 60	0.2	1.6	29.0	67.2
Bolangir	–60 + 150	0.50	3.20	12.4	83.9
	–30 + 60	0.4	2.0	9.0	88.6
TamilNadu	–60 + 150	0.23	1.77	4.20	93.8
	–30 + 60	0.2	1.1	2.5	96.2

4.2.1 *Moisture*

The moisture content of graphite in the three samples are quite low and it is good for application in refractory industry. Higher moisture leads to hydration of bricks and deteriorates the properties.

4.2.2 *Volatile Matter*

Volatile matter in all the three samples are on higher side. These volatiles include lighter aromatic hydrocarbons and other chemicals. Higher the volatiles, poorer the quality as it involves the loss on heating, in refractory industry it means the increase of porosity in the structure and deteriorates the corrosion resistance property of the brick.

4.2.3 *Ash Content*

Though ash content is quite low with Tamilnadu graphite, it is quite high with Bolangir and Palamau graphite and they are composed mainly of silica, alumina, iron oxides. In Palamau graphite, an appreciable amount of alkali oxides are also present, while Tamilnadu graphite contains increase amount of sulphur. Bolangir graphite contains higher amount of alkaline earth metals.

4.3 Particle Size Analysis

The finer fractions of samples i.e. –60BSS collected for the investigation was taken for particle size analysis. It was found that Palamau and Bolangir graphite were having almost the same particle size and finer and coarser fraction are present in equal proportions in these two samples, while in Tamilnadu grade, finer percentage is less while coarse fractions are present in substantial amount. Almost fifty percent of the samples are in the range of 40 to 100 μm. Average particle size of Bolangir and Palamau grade are almost same while that of Tamilnadu grade, average particle size is double of the other two.

4.4 Specific Gravity

Specific gravity of three samples were carried out as per ASTM standard. The result of the analysis is given in Table 4.2. It was found that specific gravity is lowest with Palamau graphite sample, mainly because of higher impurities.

Table 4.2: Specific Gravity of Three Sources of Graphite

Sample No.	*Specific Grade*
Palamau (PG)	2.14
Bolangir (BG)	2.20
TamilNadu (TG)	2.25

4.5 Specific Surface Area

Details of the result conducted by BET's method given in Table 4.3. Surface area of Palamau graphite is quite high and indicates graphite grain should be highly reactive. Bolangir graphite having the lowest surface area and activity of grains should be less in comparison to Palamau graphite. Tamilnadu graphite having highest surface area and should be most reactive among all the three sources.

Table 4.3: Specific Surface Area of Graphite Samples

Sample.	*Wt. of Sample in gm*	*Surface Area*	*Specific Surface Area m^2/gm*
Palamau	2.758	3	1.1
Bolangir	2.422	1	0.4
Tamil Nadu	1.453	2	1.4

4.6 Chemical Analysis of Ash

Chemical analysis of the graphite samples are shown in Table 4.4. The ash is found basically siliceous material with significant amount of oxides of aluminium, iron, etc. Some heavy metal oxides are also present. In Palamau graphite alkali oxides are present in quite significant amount and it affects the refractory properties of bricks also. As mentioned in literature review, presence of sulphur bearing compounds are proven to be inhibitor of oxidation reaction, it is apparent from the analysis, that Bolangir and Tamilnadu graphite are having higher sulphur content and better oxidation resistant. Presence of potassium in graphite, particularly with Palamau, also helps in faster oxidation of graphite as reported in literature by Alam and Debroy.[(113)]

Table 4.4: Chemical Analysis of Graphite Ash

Constituents, wt (%)	*Weight (%) of Ash*		
	Palamau	*Bolangir*	*Tamilnadu*
SiO_2	64.52	50.54	67.54
Al_2O_3	23.85	33.42	14.67
Fe_2O_3	4.87	11.33	13.52
CaO	0.45	0.84	0.87
MgO	1.08	1.44	0.96
TiO_2	0.35	0.90	0.37
Na_2O	0.17	0.21	0.29
K_2O	3.93	0.41	0.46
V_2O_5	0.16	0.04	0.085
MnO	0.021	0.057	0.044
NiO	0.017	0.031	0.065
CuO	0.043	0.044	0.024
ZnO	0.025	0.014	0.022
ZrO_2	0.033	0.015	0.09
Cir_2O_3	0.31	0.082	0.15
P_2O_5	0.088	0.21	0.24
S	0.0057	0.34	0.45

4.7 Differential Thermal and Thermo Gravimetric Analysis

Figures 4.1–4.3 show the DTA/TGA traces of different graphite samples. All the traces show same type of behaviour having three endothermic

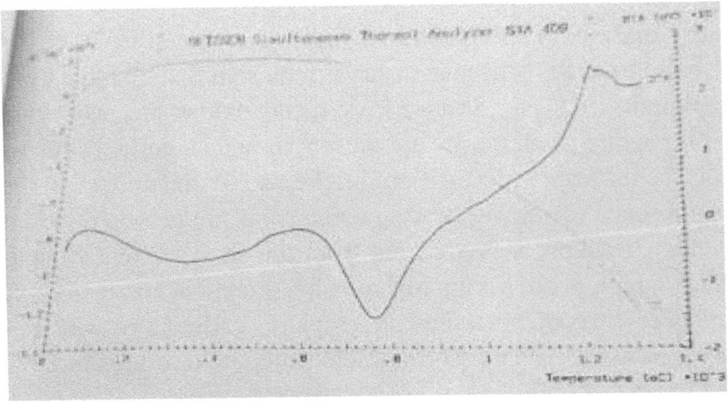

Fig. 4.1: DTA of Palamau Graphite

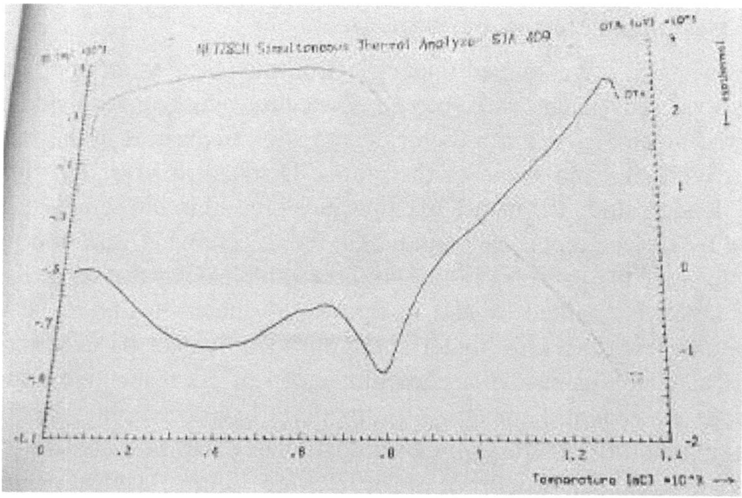

Fig. 4.2: DTA of Bolagir Graphite

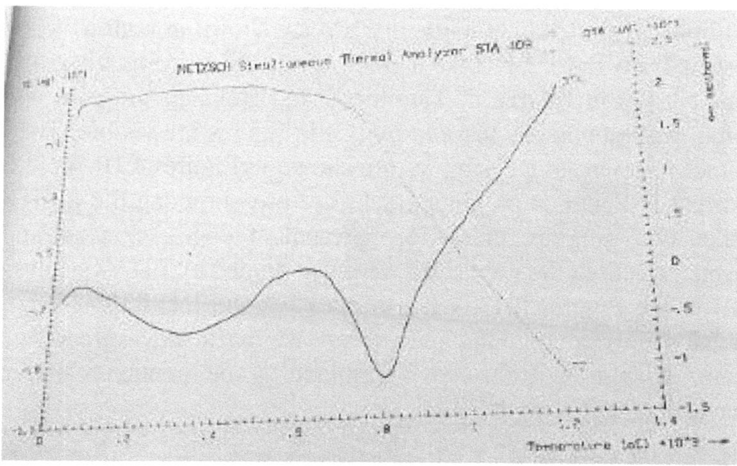

Fig. 4.3: DTA of Tamilnadu Graphite

peaks in air atmosphere. In case of Palamau graphite peak start at slightly lower temperature than the other two samples. While for Palamau, the peaks appears at 630°C and that for others, it comes at 660°C. The peak in the range of (630–900°C) comes because of oxidation of graphite powder and consequent weight loss in the samples as depicted by thermo-gravimetric curve. The weight loss nature in all the three samples are quite similar and predicts the similar behaviour of the samples. From these DTA/TGA curves, the initial starting of oxidation of graphite has been estimated.

4.8 Scanning Electron Microscopy

Three samples were studied under electron microscopy and topography, back scattered image and secondary electron image were studied to observe the cleavage, grain diameter, porosity, impurities in the samples and overall size and shape of the grains. Detailed SEM-EDAX analysis were done to study the nature of impurities present in the samples. It was found by surface topography study that there were wide variation in size and shape of the particles in all the three samples. All the samples have layer structure marked by (cl) in figures. as shown in Figures 4.4–4.6. There are few cracks in bigger size grains. Individual particle size has been calculated using the surface dimension of the respective particles. Average equivalent diameter of the particles have been obtained from a large number of randomly chosen particles. The average equivalent diameter of the Palamau grade of particles are 70 μm. Average equivalent diameter of Bolangir grade particles are 165 μm., while that of Tamilnadu grade particles are 115 μm. Secondary electron image of the Palamau, Bolangir and Tamilnadu graphite in SEM are shown in Figures 4.7–4.9. It was seen from the study that there were fairly large number of pores marked by (P) in figures in Tamilnadu and Bolangir graphite while in Palamau graphite cracks marked by (C) in figures are visible. The pores and cracks present in the samples are shown in Figures 4.10–4.12. It was found that in Palamau graphite cracks are present along the plane while in other two samples, cracks are present at edges. Concentration of elements present in the impurities were determined by EDAX analysis and shown in the Figures 4.13–4.18. It can be found that Palamau graphite contains mainly aluminium and silicon while Bolangir grade contains calcium, silicon & Iron and Tamilnadu grade contains aluminium, silicon, calcium, iron & titanium.

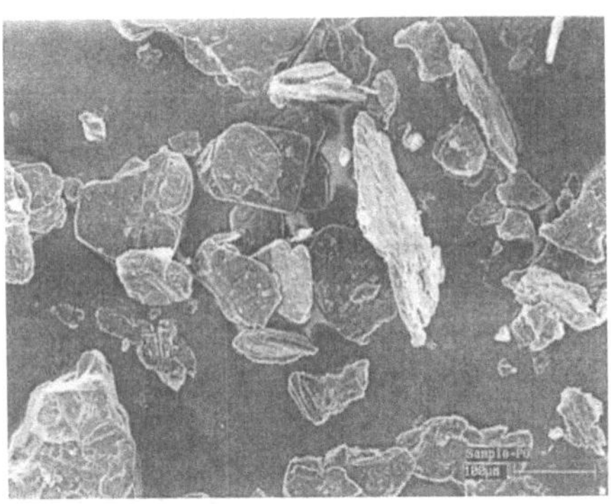

Fig. 4.4: Topographic View of Palamau Graphite in SEM

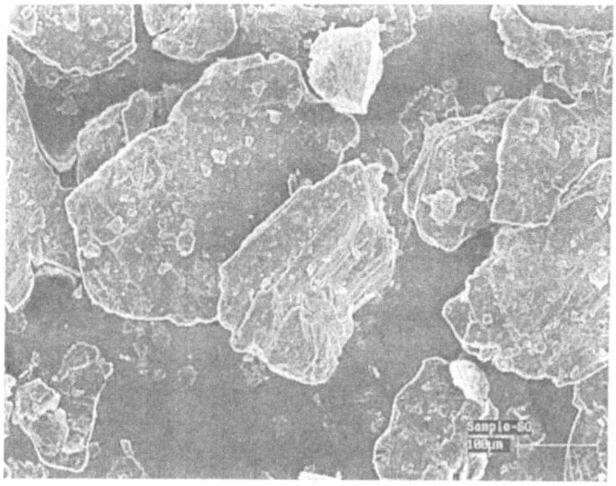

Fig. 4.5: Topographic View of Bolangir Graphite in SEM

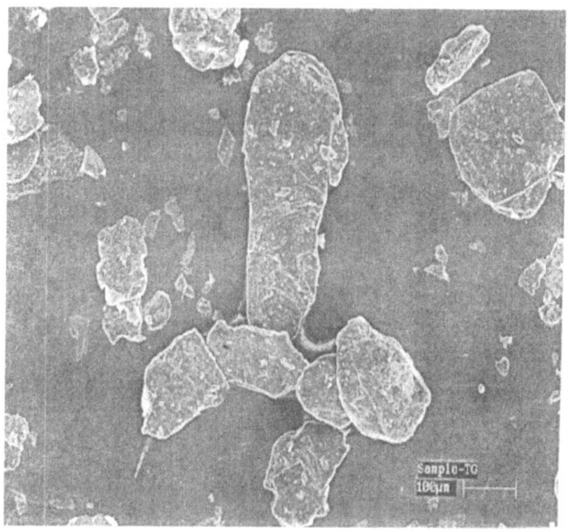

Fig. 4.6: Topographic View of Tamilnadu Graphite in SEM

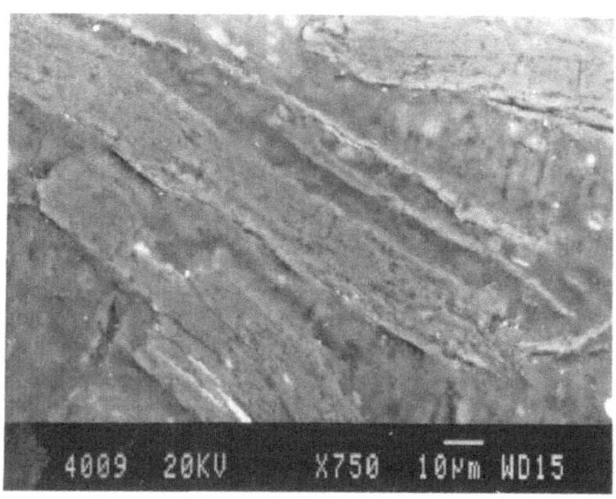

Fig. 4.7: Secondary Electron Image of Palamau Graphite in SEM

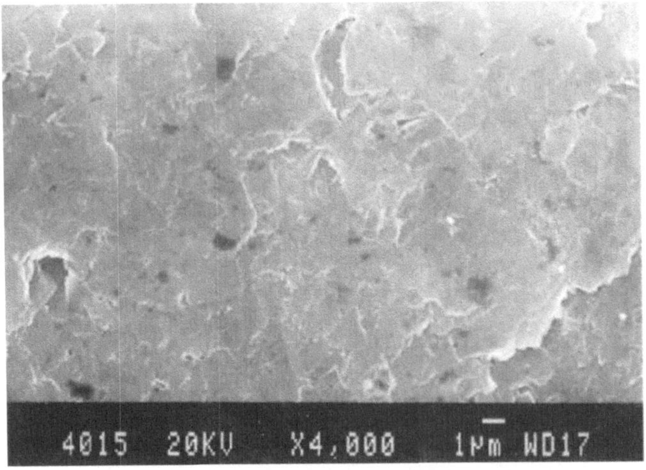

Fig. 4.8: Secondary Electron Image of Bolangir Graphite

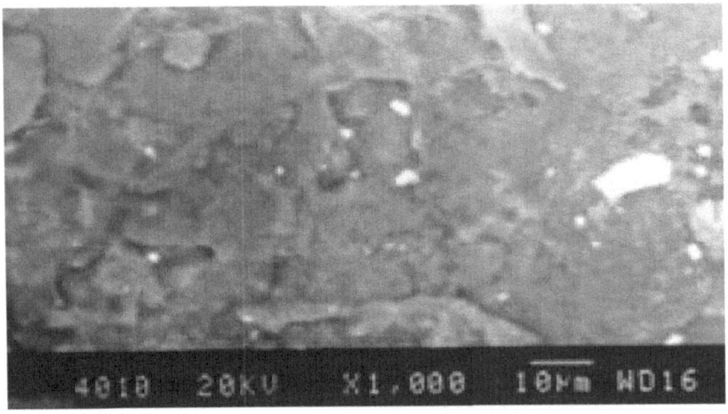

Fig. 4.9: Secondary Electron Image of Tamilnadu Graphite in SEM

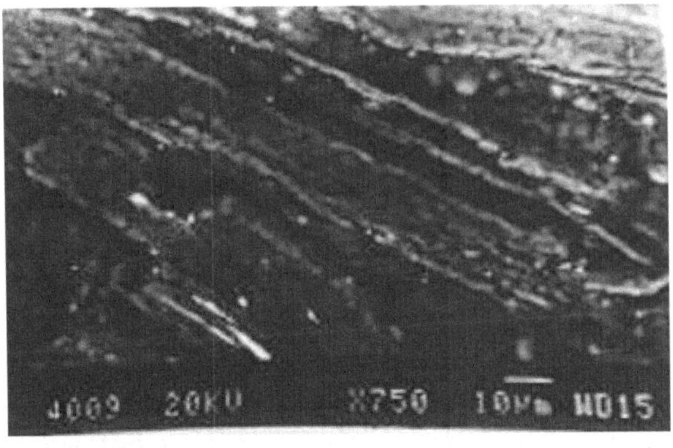

Fig. 4.10: SEM Photomicrograph of Palamau Graphite Showing Cracks and Pores in the Grain

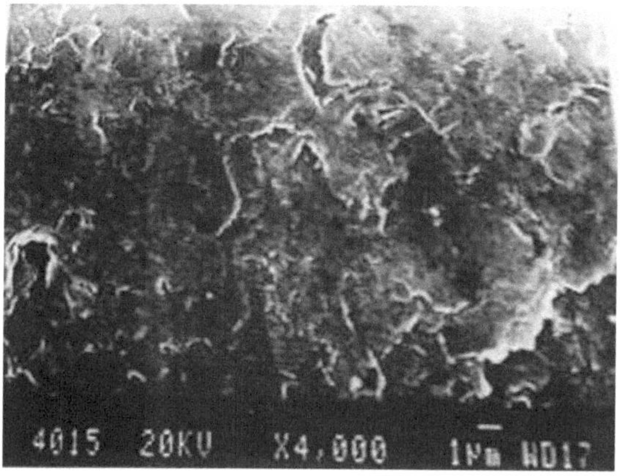

Fig. 4.11: SEM Photomicrograph of Bolangir Graphite Showing Cracks and Pores in the Grain

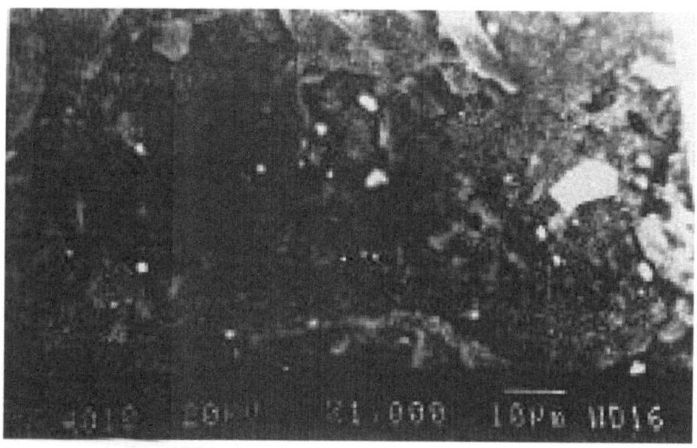

Fig. 4.12: SEM Photo Micrograph of Tamilnadu Graphite Showing Pores and Cracks in the Grain

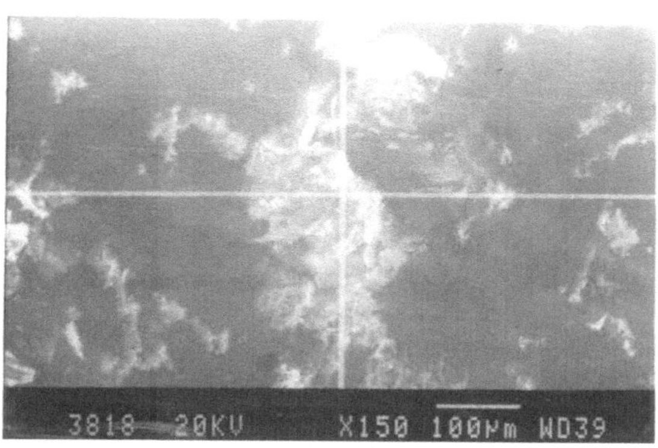

Fig. 4.13: SEM/EDAX Showing Distribution of Impurities in Palamau Graphite at X

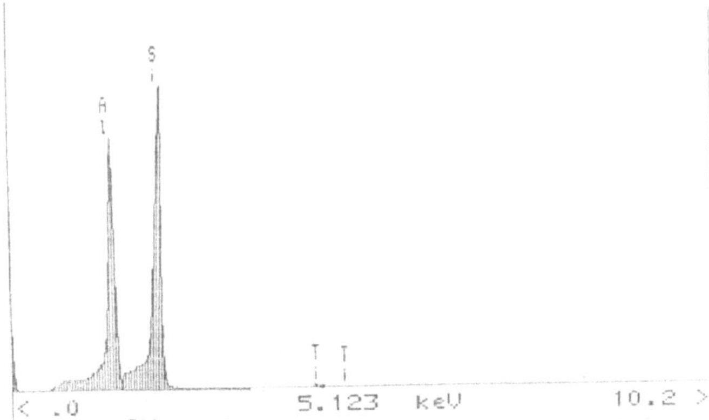

Fig. 4.14: EDAX of Above Sample at X

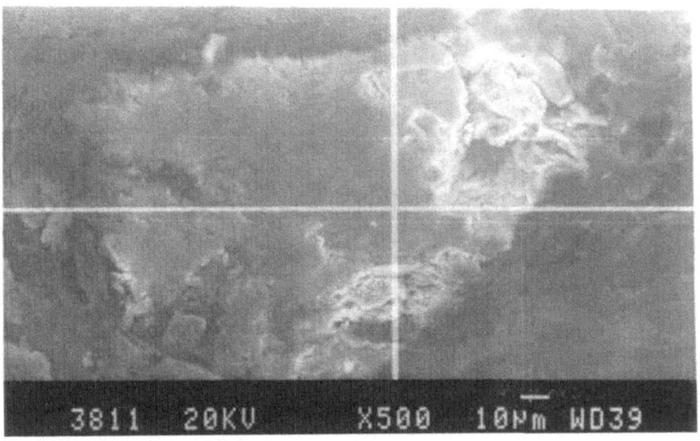

Fig. 4.15: SEM/EDAX Showing Distribution of Impurities in Bolangir Graphite at X

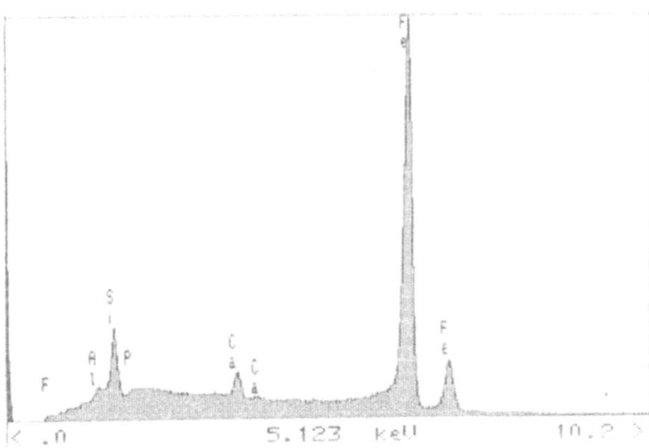

Fig. 4.16: EDAX of Above Sample at X

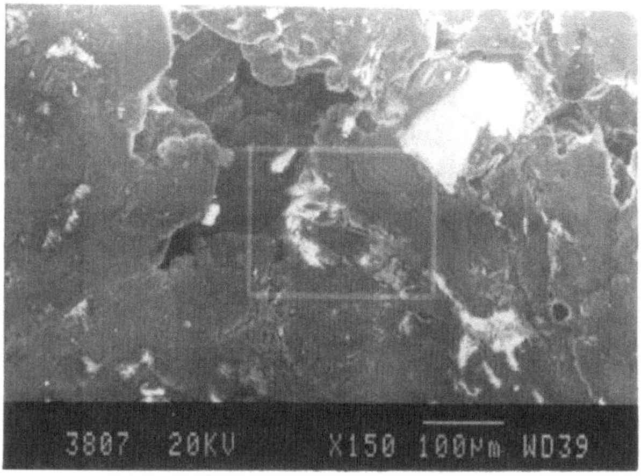

Fig. 4.17: SEM/EDAX Showing Distribution of Impurities in Tamilnadu Graphite at X

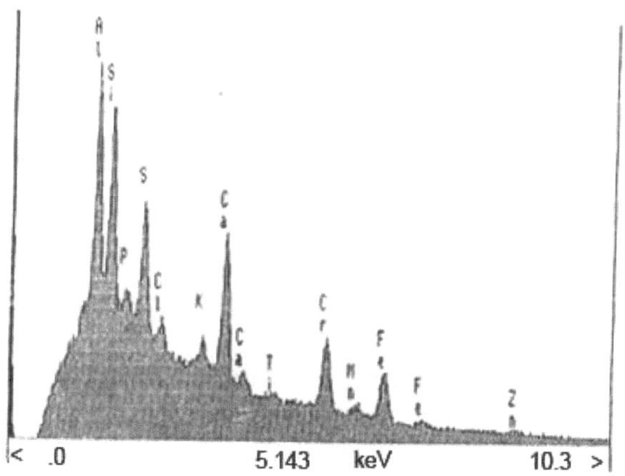

Fig. 4.18: EDAX of Above Sample at X

4.9 X-ray Diffraction Analysis

X-ray diffraction study conducted of the three samples in BRUKER-AXS, model No. DN8 Advance. The patters obtained are given in Figures 4.19–4.21. From the figures, it is apparent that all the samples are highly crystalline and peaks obtained are quite sharp. Both (002) & (004) reflections of the tested samples are quite sharp. The interlayer distance of the three samples were calculated from the average of d(002) & d(004) values. It is seen that interlayer distance are very close and the corresponds to the values of pure graphite structure. It is also seen from the

XRD-analysis that no other phases are present in the XRD curves which indicates that the impurity percentage in the sample are very less and though they are present quite substantially in Palamau and Bolangir region of graphite but individual compounds present in the samples are very less, and that causes the absence of peaks in the XRD-curves. The degree of graphitisation (g) was calculated for all three samples and the results including the inter layer spacing values are presented in Table 4.5. It is apparent from the results that all the samples are highly graphitised and in Palamau graphite, degree of grahitisation is less i.e. 85.0%.

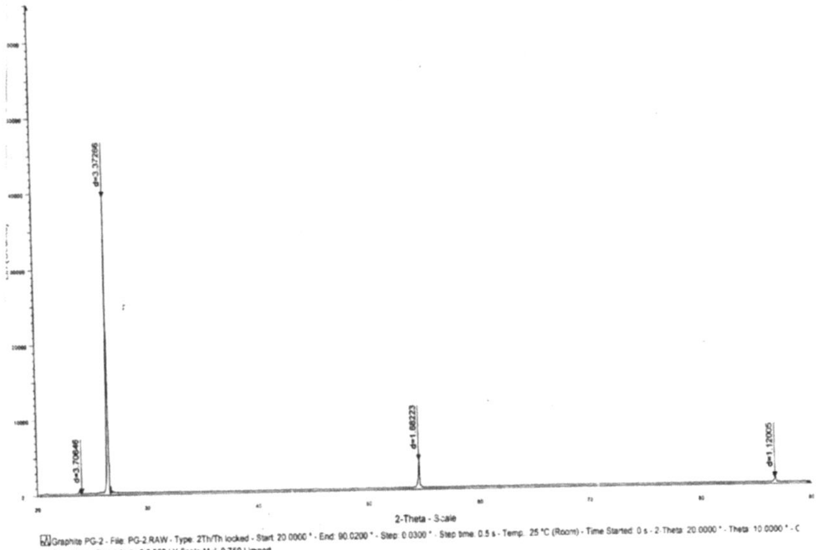

Fig. 4.19: XRD Chart of Palamaur Graphite

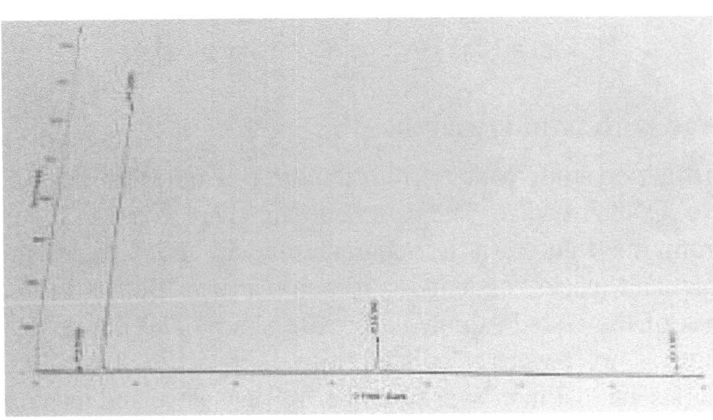

Fig. 4.20: XRD Chart of Bolangir Graphite

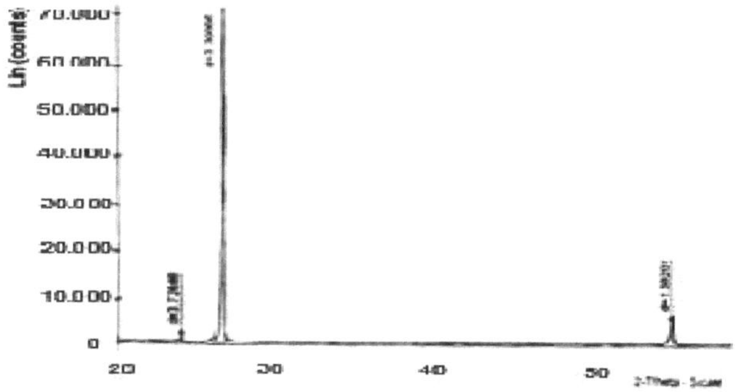

Fig. 4.21: XRD Chart of Tamilnadu Graphite

From the position of the (002) line, it was found in this study that the d-spacing in all the three samples, were greater than the characteristics value of 3.354 Å for purely crystalline graphite which shows that some non-crystalline graphite are also present in the samples and that is also confirmed while calculating the degree of grahitisation as shown in Table 4.5.

Table 4.5: Degree of Graphitisation in the Graphite Samples

Sample No.	*D(002) in A°*	*D(004) in A°*	*C in A°*	*G*
Palamau	3.350	3.359	6.709	0.85
Bolangir	3.366	3.367	6.733	0.98
Tamil Nadu	3.356	3.365	6.721	0.87

4.10 Petrographic Analysis

All the three graphite samples were characterized petrograhically. Details of micrographic studies were carried out under transmitted light. Palamau graphite was jet black to steel grey coloured, medium to fine grained. Grains were flaky and have well developed basal cleavage and metallic lusture. Grains were crystalline. Columnar grains marked by (CO) in Figure and radiated crystals were also found. Most of the grains were basal flakes. Figure 4.22(a)&(b) (both under plane and cross polars and under the magnification of 10X) shows the character. Bolangir graphite was steel grey, flaky, medium to coarse grained. All the grains were crystalline. Rhombohedral patterns of grains marked by (RH) in figure were also present. Few columnar and radiated crystals were also found. Graphite shows well developed basal cleavage. Figure 4.23(a)&(b) (both under plane and cross polars and under the magnification of 10X) shows the character. Tamilnadu graphite was black to steel grey in colour. Most of the grains were medium to fine Few coarse grained minerals were also common Most of the grains show columnar and radiated pattern, others were rhombohedral shown in Figure by (CO) and (RH). The mineral shows well developed cleavage. Figure 4.24(a)&(b) (both under plane and cross polars and under the magnification of 10X shows the character.

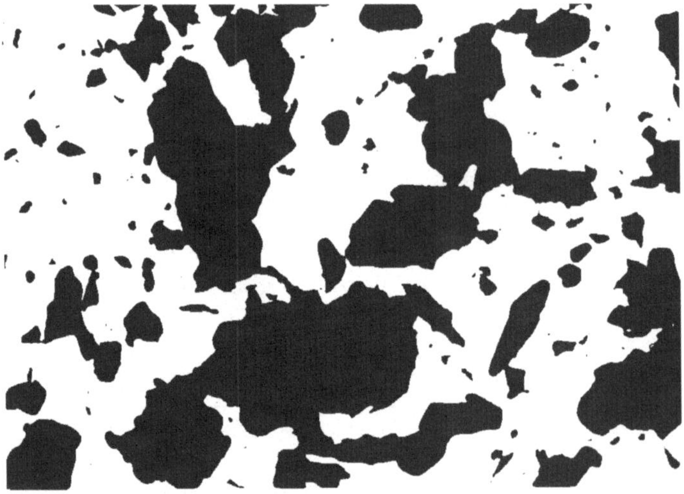

Fig. 4.22(a): Photomicrograph of Pamalau Graphite (plane polar), 10X

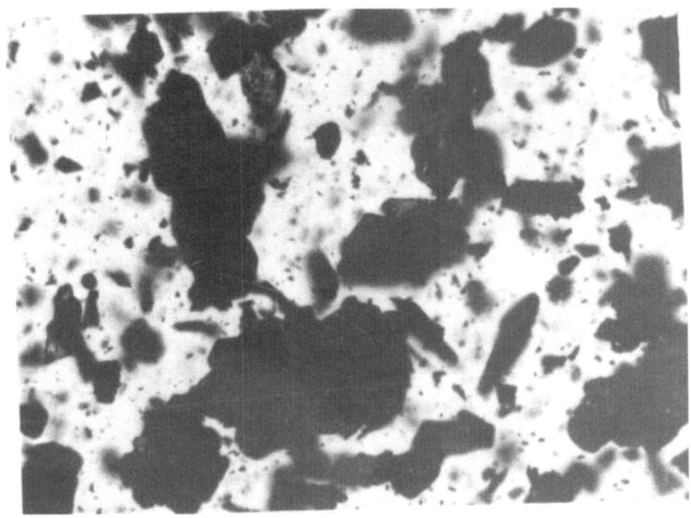

Fig. 4.22(b): Photomicrograph of Pamalau Graphite (cross polar), 10X

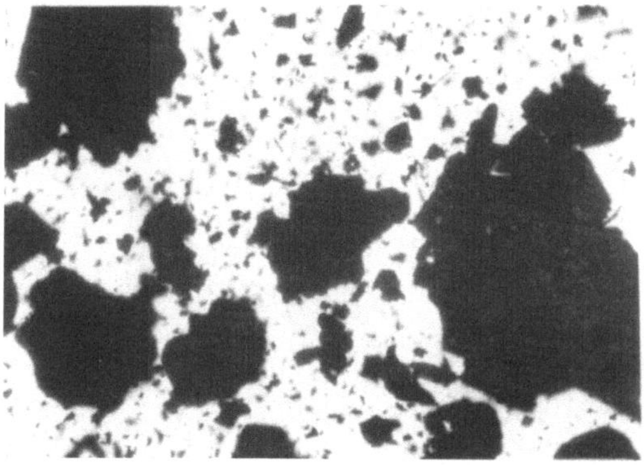

Fig. 4.23(a): Photomicrograph of Bolangir Graphite (plane polar), 10X

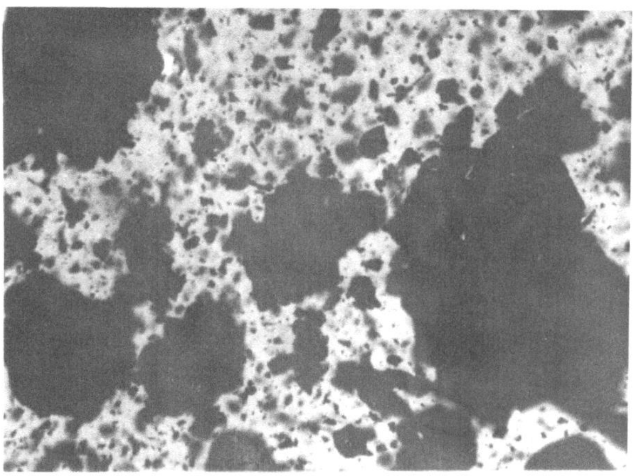

Fig. 4.23(b): Photomicrograph of Bolangir Graphite (cross polar), 10X

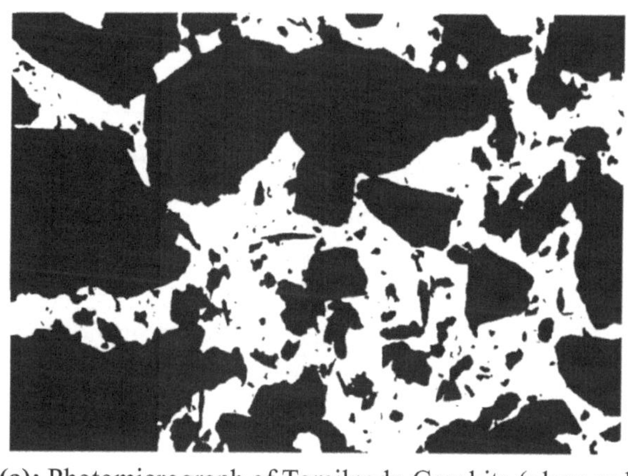

Fig. 4.24(a): Photomicrograph of Tamilnadu Graphite (plane polar), 10X

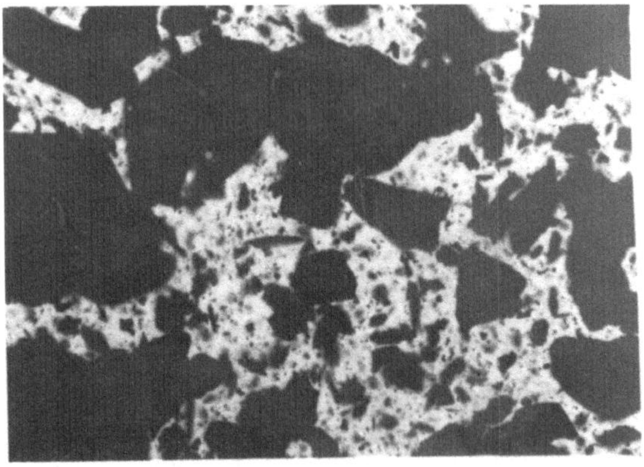

Fig. 4.24(b): Photomicrograph of Tamilnadu Graphite (cross polar), 10X

4.11 Graphite Ash Fusion Temperature

Three samples were tested under heating microscope to study the nature of impurities with rising temperature. It has been found that in Palamau graphite initial starting temperature of fusion is lower compared to the other two. Detailed test results are given in Table 4.6. Lower temperature of Palamau grade is because of the higher alkali content in the sample. However, softening temperature of Palamau graphite is much higher than the other two. This may be due to the presence of the higher amount of alumina and silica in the ash sample. In both Bolangir and Tamilnadu grade softening temperature is same almost. This may be due to the presence of similar kind of ingredients present in both the ash samples. Details of the ingredients present in the three ash sample were analysed by XRD of ash samples and given in Figures 4.25–4.27.

Table 4.6: Deformation Temperature and Phase Identification of Ash in Graphite Samples

Sample No.	*Initial Temp °C, of Deformation*	*Phase Identification*
Palamau	1075	Quartz, feldspar, corundum, potassium-iron-magnesium-aluminium-silicate hydroxide-hydrate
Bolangir	1125	Quartz, feldspar, pyrite, nontronite clay, corundum, calcium-magnesium-aluminium-silicate- maliss clay.
Tamil Nadu	1105	Quartz, feldspar, corundum, pyrite, sodium-iron-aluminium silicate hydrate

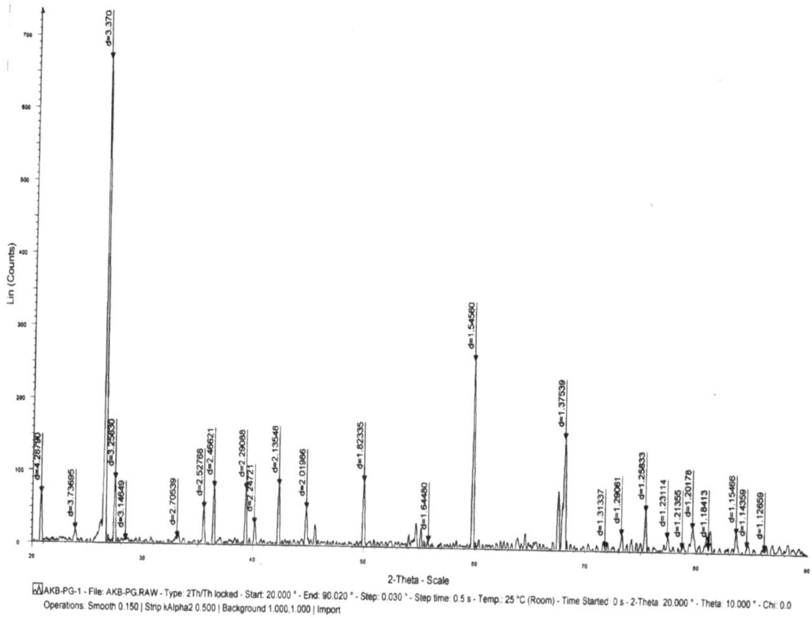

Fig 4.25: X-ray Diffraction Pattern of Ash of Palamau Graphite

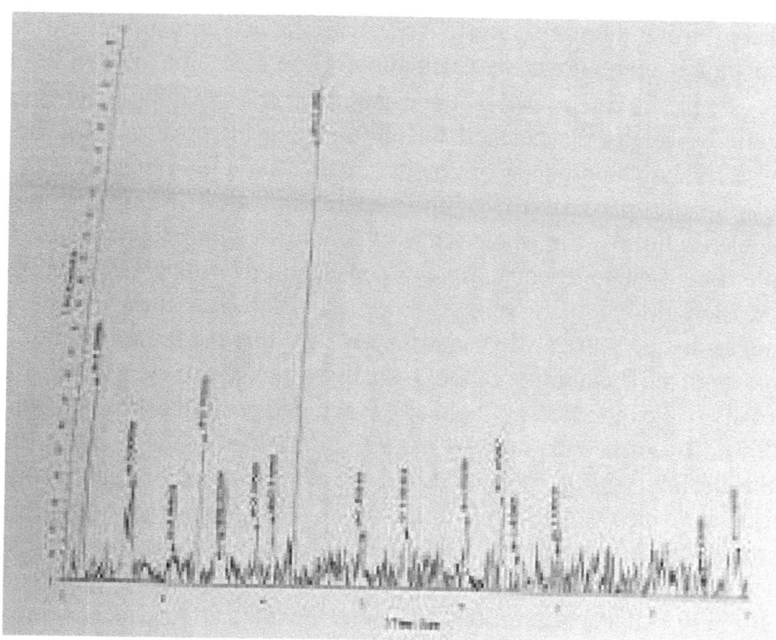

Fig. 4.26: XRD Curve of Bolangir Graphite Ash Sample for Phase Identification

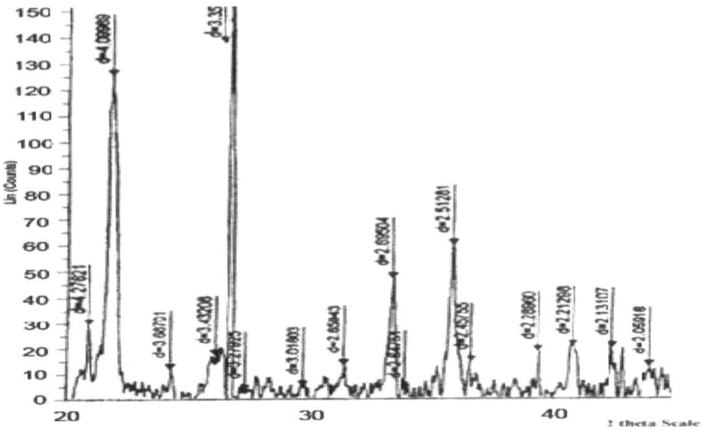

Fig. 4.27: XRD Curve of Tamilnadu Graphite Ash Sample for Phase Identification

4.12 Oxidation Rate Study

The information collected in this sort of experiments are α, the fraction of original reactant decomposed against time '*t*'. Such a plot is automatically recorded in DT/TG analysis and numerous literature deal with information gathered from these reactions. Repeating the experiments at different temperatures then allows the calculation of the activation energy. T form of α vs. *t* plot is determined by the changing geometry of the decomposing system, especially the geometry of the reacting interface which changes progressively with progress of time because of variations in particle shape, size, porosity, permeability of the system, etc. As a result, in a homogenous reaction[117] the whole range of decomposition fits one kinetic law while in a heterogeneous system consisting of solids or even forth expression during increasing time period. Moreover these progressive changes of stages are always interlapped such that the termination of one stage and the beginning of next change are always overlapping and porosity one expression my hold a few percentage of decomposition to be followed by a second, third in nature. Difficulties are also experienced while dealing with porous solid and the phenomenon of "cannibalism" effect, in which small pores merging together with the bigger pores associated with reduction of porosity but increase of pore diameter taking place may find enhanced avenues for escape of the gaseous product, i.e, CO_2. Phase boundary controlled reactions in which an interface is moving with constant velocity was indicative of chemical reaction between carbon and oxygen or air and describes low temperature oxidation.

CHAPTER 5

RESULTS AND ANALYSIS

5.1 Palamau Graphite

5.1.1 *Chemical Analysis of Ash*

Done through X-ray fluorescence method as given in Table 4.4.

5.1.2 *Physical Analysis*

5.1.2.1 *Proximate Analysis*

Details given in Table 4.1.

5.1.2.2 *Particle Size Analysis*

Average particle size and grain size distributed in (–60) fines is indicated below.

Average particle size (done through SEM) – 70 μm

Grain size Ranges – done through LASER diffraction method:

2–5 μ	-	12%
5–10 μ	-	14%
10–20 μ	-	23%
20–30 μ	-	10%
30–50 μ	-	31%
50–100 μ	-	10%

In addition SEM image through secondary electron mode indicates large amount of pores and cracks in the grains as shown in Figure 4.10.

5.1.2.3 *Grain Density*

This was done by Pycnometer method (using xylene as immersion liquid) while tap density and loose density were done by tapping a fixed amount of the graphite for 40 times inside a cylinder and filling the graphite sample loosely inside a cylinder having fixed graduations respectively. The figures obtained are as follows.

Grain density	:	2.14 gm/cc
Tap density	:	3.34 gm/cc
Loose density	:	2.94 gm/cc

5.1.2.4 *Surface Area*

It was determined by BET method indicates average surface area as 1.1 m^2/gm.

5.1.2.5 *Fusion Temperature of Ash*

It is known that the impurities present in the sample as shown in SEM/EDAX analysis, affects the oxidation behaviour significantly, a detailed ash analysis present in the graphite sample after complete ignition of the graphite is indicated below.

Fusion temperature of ash – 1075°C
Hemispherical shape formation temperature >1400°C

X-ray diffraction analysis of ash sample shows the presence of quartz, corundum, feldspar and alkali-aluminium-iron-magnesium-silicate hydrate complex.

5.1.2.6 *Degree of Gaphitisation*

As done by x-ray diffraction method and then using Maire & Merings formula, indicates 85% degree of graphistisation and graphitised phase present in the graphite is 67% as identified by XRD.

5.1.2.7 *Oxidation Behaviour*

The extent of oxidation of graphite sample as a function of time was followed gravimetrically and isothermally at temperatures from 735°C to 830°C in both air and oxygen atmosphere in DTA/TGA apparatus. The temperatures was monitored by means of a platinum/rhodium thermocouple located directly below the specimen.

Figures 5.1–5.4 show the plots of fraction of graphite (α) oxidized against time at various temperatures with two fractions of graphite in oxygen and air atmosphere. Flow rate of air and oxygen (i.e. 0.25 bar) inside the furnace was kept constant by means of flow regulator. The rate decreased with extent of reaction at any temperature and the effect of temperatures, atmosphere and grain size is evident. At higher temperature, the α-time curve is very sharp and shape of the α-time plots at different temperatures are different.

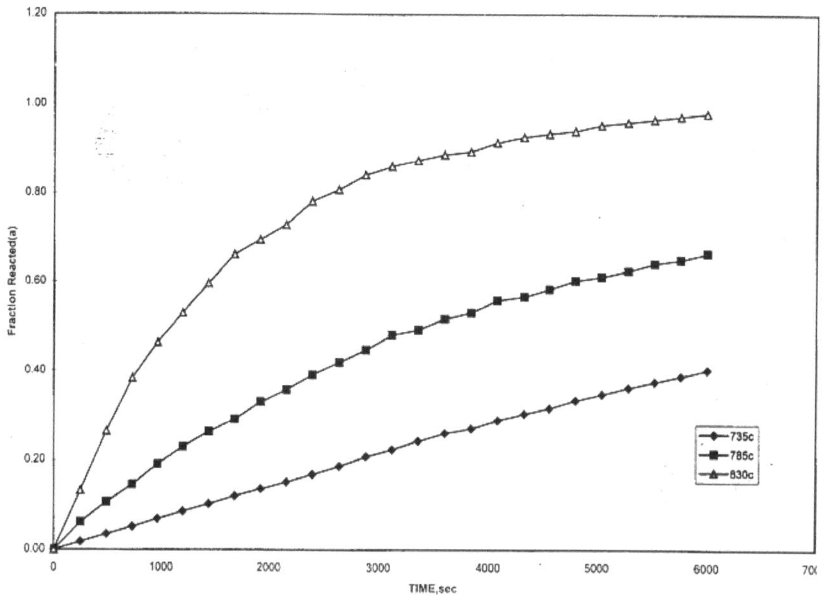

Fig. 5.1: Fraction Graphite Reacted as a Function of Time-PG1 (air)

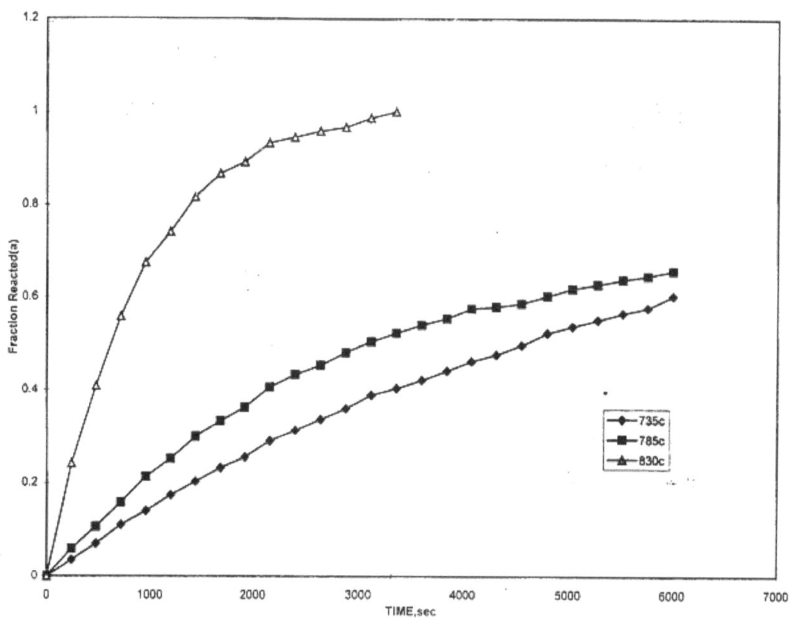

Fig. 5.2: Fraction Graphite Reacted as a Function of Time-PG1 (oxygen)

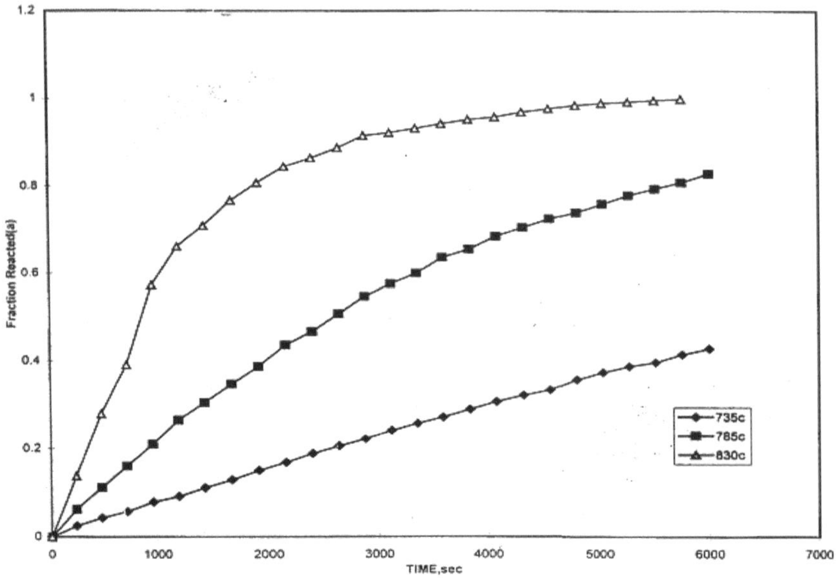

Fig. 5.3: Fraction Graphite Reacted as a Function of Time-PG2 (air)

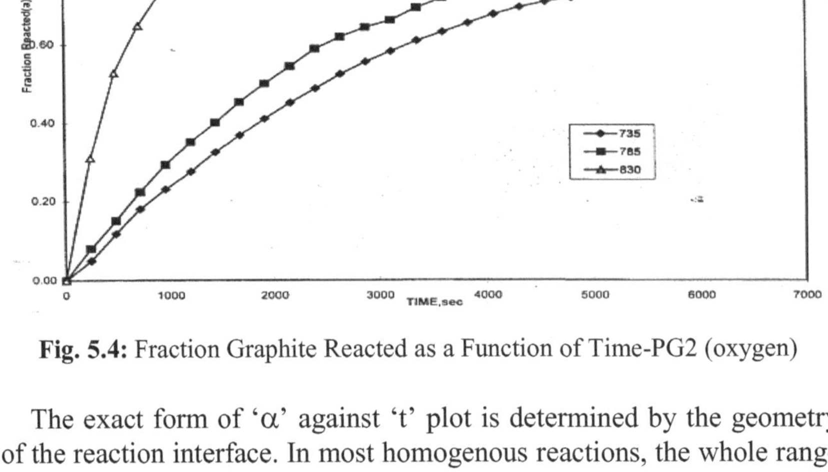

Fig. 5.4: Fraction Graphite Reacted as a Function of Time-PG2 (oxygen)

The exact form of 'α' against 't' plot is determined by the geometry of the reaction interface. In most homogenous reactions, the whole range of decomposition fits one kinetic law but in heterogeneous reactions, one expression may hold for the initial few percent of the reaction to be followed by a second in the accelerating stage of the reaction and a third or fourth expression may be found to be described the reaction in the decelerating stage of reaction.

Reaction of porous carbon solids in oxidizing gases frequently shows three temperature regimes.[50,119] At low temperature rate is controlled by the reaction at the interface, the rate of diffusion of reacting gaseous species to the internal surfaces of the solid being rapid, so that no concentration gradient exists within the pore structure, which is uniformly eroded. As the temperature increases, the rate of reaction first becomes comparable with the rate of diffusion through the pore structure so that the concentration of reactant falls to zero within the pore structure and then the reaction at higher temperature is controlled by mass transport of reactant and product gases across the boundary layer at the external surface of the porous solid. These processes have been discussed previously.[81] Turkdogan *et al.*[82] concluded that depending on the size, porosity and CO_2 pressure, the rate of oxidation of graphite is controlled by the chemical reaction upto about 1100°C. Above this temperature mass transfer control takes place.

5.1.3 *Summary of Observations*

- It has been found that grain size does not have impact on fraction decomposed for both size fractions. From the (α-t) curves, it is apparent that curves for two fractions in both ranges are superimposed, which suggest that oxidation of graphite is independent of size fractions of (–30+150 BSS) and such phenomenon has been reported by other authors.[85,88]
- Palamau graphite used here contains higher amount of ash i.e, 18% with higher amount of alkali i.e., 3.5%.
- Degree of graphitsiation is high i.e, 85% and percent of graphitised phase present in the sample is 67%.
- Ash contains impurities like quartz, feldspar, corundum, alkali-magnisium-aluminimum-silicate hydrates.
- Large amount of cracks are present in the sample as depicted by SEM.

5.2 Bolangir Graphite

5.2.1 *Chemical Analysis of Ash*

Done through X-ray fluorescence method given in Table 4.4.

5.2.2 *Physical Analysis*

5.2.2.1 *Proximate Analysis*

Details are given in Table 4.1.

5.2.2.2 *Particle Size*

Average particle size was done through SEM and found to be 165 μm.

Size ranges of –60BSS fine grains done through LASER diffraction method.

2–5 μ	–	11%
5–10 μ	–	16%
10–20 μ	–	21%
20–30 μ	–	13%
30–50 μ	–	25%
50–100 μ	–	14%
> 100 μ	–	NIL

In addition SEM image through secondary electron mode indicates large amount of pores in the grain as show in Figure 4.11.

5.2.2.3 *Grain Density*

Grain density was done by Pycnometer method (using xylene as immersion liquid) while tap density and loose density were done by tapping a fixed amount of the graphite for 40 times inside a cylinder and filling the graphite sample loosely inside a cylinder having fixed graduation respectively. The values obtained were as follows:

Grain density	:	2.20 gm/cc
Tap density	:	3.8 gm/cc
Loose density	:	2.24 gm/cc

5.2.2.4 *Surface Area*

Determined by BET method indicates average surface area as 0.4 m^2/gm.

5.2.2.5 *Fusion Temperature of Ash*

It is known that the impurities present in the sample as shown in SEM/EDAX analysis, affects the oxidation behaviour significantly, a detailed ash analysis present in the graphite sample after complete ignition of the graphite is indicated below.

Fusion temperature of ash – 1125°C

Hemispherical shape formation temperature –1290°C.

X-ray diffraction analysis of ash sample shows the presence of quartz, feldspar, corundum alongwith calcium-magnesium-aluminium silicate hydrate and nontronite clay.

5.2.2.6 *Degree of Graphitisation*

It was done by the x-ray diffraction method and then using Maire & Merings formula, indicates 98% graphitisation and graphitised phase present in the graphite was 87%.

5 2.2.7 *Oxidation Rate Study*

The extent of oxidation of graphite sample as a function of time was followed gravimetrically and isothermally at temperatures from 735°C to 830°C in both air and oxygen atmosphere in DTA/TGA apparatus. Temperature was monitored by means of a platinum/rhodium thermo-couple located directly below the specimen.

Figures 5.5–5.8 shows the plots of fraction of graphite (α) oxidized against time at various temperatures with two fractions of graphite and in

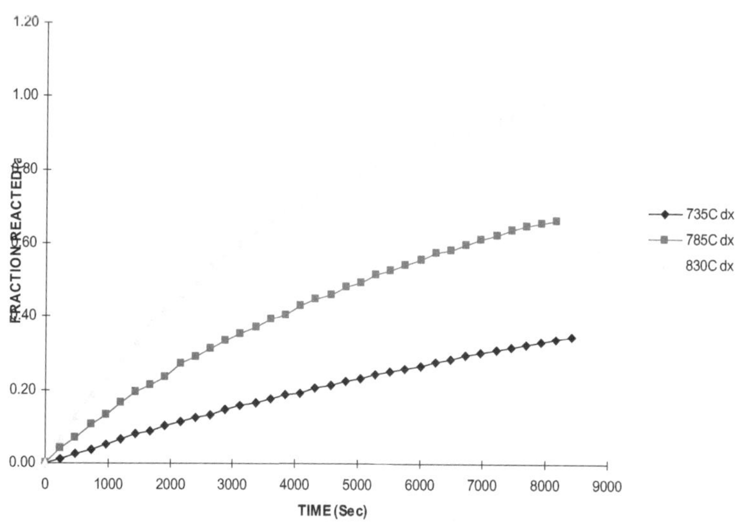

Fig. 5.5: Fraction Graphite Reacted as a Function of Time (SG1,Air)

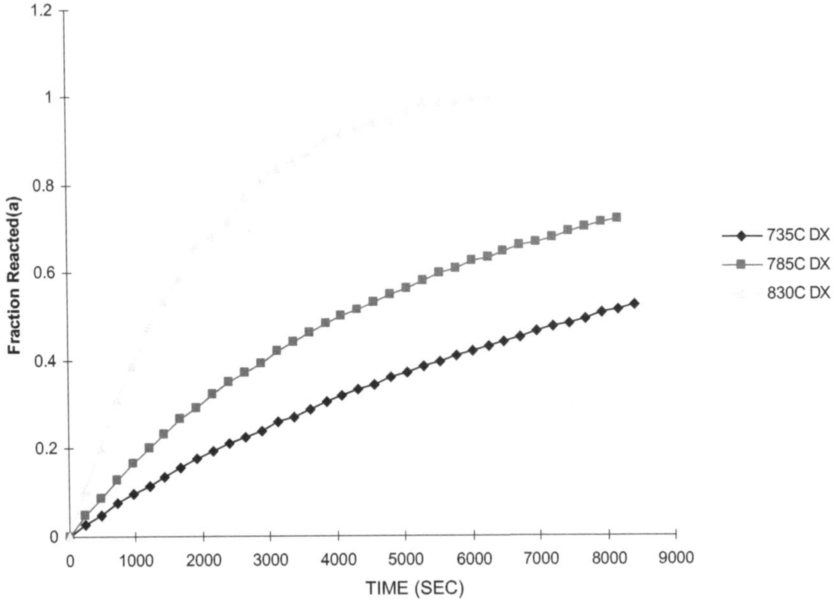

Fig. 5.6: Fraction Graphite Reacted as a Function of Time (SG1, Oxygen)

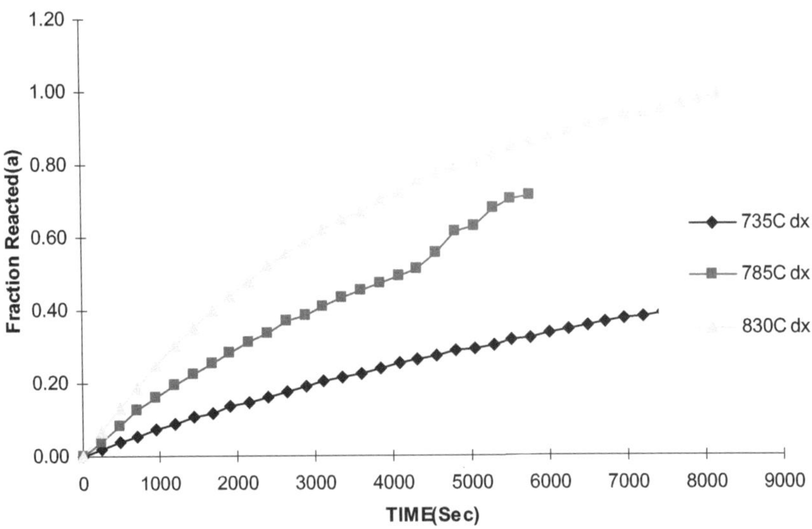

Fig. 5.7: Fraction Graphite Reacted as a Function of Time (SG2, Air)

oxygen and air atmosphere. Flow rate of air and oxygen (i.e. 0.25 bar) inside the furnace was kept constant by means of flow regulator. The rate decreased with extent of reaction at any temperature and the effect of temperatures, atmosphere and grain size is evident. At higher temperature, the α-time curves were very sharp and shape of the α-time plots at different temperatures were different.

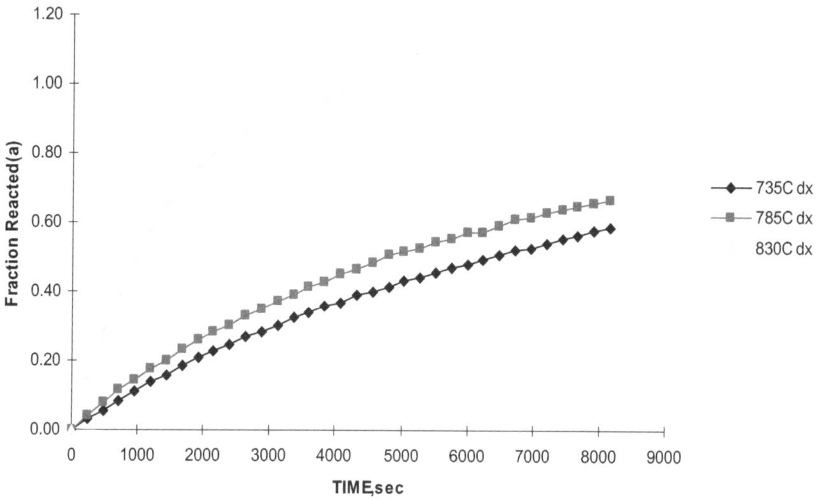

Fig. 5.8: Fraction Graphite Reacted as a Function of Time (SG2, Oxygen)

The exact form of 'α' against 't' plot is determined by the geometry of the reaction interface. In most homogenous reactions, the whole range of decomposition fits one kinetic law but in heterogeneous reactions, one expression may hold for the initial few percent of the reaction to be followed by a second in the accelerating stage of the reaction and a third or fourth expression may be found to be describe the reaction in the decelerating stage of reaction.

5.2.3 *Summary of Observations*

- It has been found that grain size does not have impact on fraction decomposed for both size fractions. From the (α-t) curves, it is apparent that curves for two fractions in air and oxygen atmosphere are superimposed, which suggest that oxidation of graphite is independent of size fractions of (–30+150 BSS) and such phenomenon has been reported by other authors.[85,88]
- Bolangir graphite used here contains higher amount of ash i.e., 12% and surface area is 0.4 m^2/gm.

- Degree of graphitsiation is very high i.e., 98% and percent of graphitised phase present in the sample is 87%.
- Ash contains impurities like quartz, feldspar, corundum, calcium-magnesium-aluminimum-silicate hydrate, nontronite clay etc.
- Large amount of pores are present in the sample as depicted by SEM.

5.3 Tamilnadu Graphite

5.3.1 *Chemical Analysis of Ash*

Done through Xray fluorescence method is given in Table 4.4.

5.3.2 *Physical Analysis*

5.3.2.1 *Proximate Analysis*

Given in Table 4.1.

5.3.2.2 *Particle Size*

Average particle size analysis and the grain size distribution is indicated below.

Average particle size (done through SEM) – 115 μm.

Size ranges of (–60) BSS fine grains(done through LASER diffraction method).

2.5 μm	-	4.5%
5–10 μm	-	5.5%
10–20 μm	-	10%
20–30 μm	-	11%
30–50 μm	-	23%
50–100 μm	-	35%
>100 μm	-	11%

In addition SEM usage through secondary electron mode indicates large amount of pores in the particles as shown in Figure 4.12.

5.3.2.3 *Grain Density*

It was done by pycnometre method (using xylene as immersion liquid) while tap and loose density was done by tapping a fixed amount of the graphite for 40 times inside a cylinder and filling the graphite sample loosely inside a cylinder having fixed graduations respectively. The values obtained were grain density 2.25 gm/cc, tap density 3.2 gm/cc and loose tap density 2.6 gms/cc

5.3.2.4 *Surface Area*

It was determined by BET method which indicates average surface of 1.4 m^2/gm.

5.3.2.5 *Fusion Point of Ash*

It is known that the impurities present in the sample as shown in SEM& EDAX analysis affects the oxidation behaviour significantly. A detailed ash analysis present in the graphite after complete ignition of the graphite is indicated below.

Fusion temperature of ash – 1105°C

Hemispherical shape formation temperature – 1280°C

X-ray diffraction analysis shows mainly quartz, feldspar, alongwith complexes like alkali-iron-aluminium silicates and pyrites etc. phases in graphite.

5.3.2.6 *Degree of Graphitisation*

It was done by the method of x-ray diffraction and using Mair & Merings formula indicates 87.2% graphitization and graphitised phase present in the sample is 93%.

5.3.2.7 *Oxidation Rate Study*

The extent of oxidation of graphite sample as a function of time was followed gravimetrically and isothermally at temperatures from 735°C to 830°C in both air and oxygen atmosphere in DTA/TGA apparatus. Temperature was monitored by means of a platinum/rhodium thermo-couple located directly below the specimen.

Figures 5.9–5.12 shows the plots of fraction of graphite (α) oxidized against time at various temperatures with two fractions of graphite and in oxygen and air atmosphere. Flow rate of air and oxygen (i.e. 0.25 bar) inside the furnace was kept constant by means of flow regulator. The rate decreased with extent of reaction at any temperature and the effect of temperatures, atmosphere and grain size is evident. At higher temperature, the α-time curves were very sharp and shape of the α-time plots at different temperatures were different.

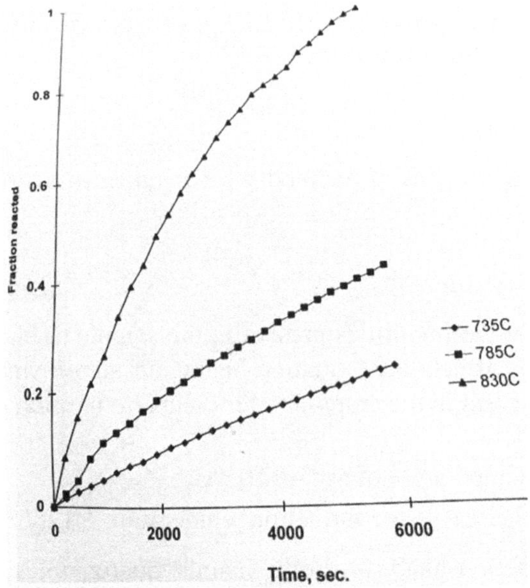

Fig. 5.9: Fraction Graphite Reacted as a Function of Time TG1 (air)

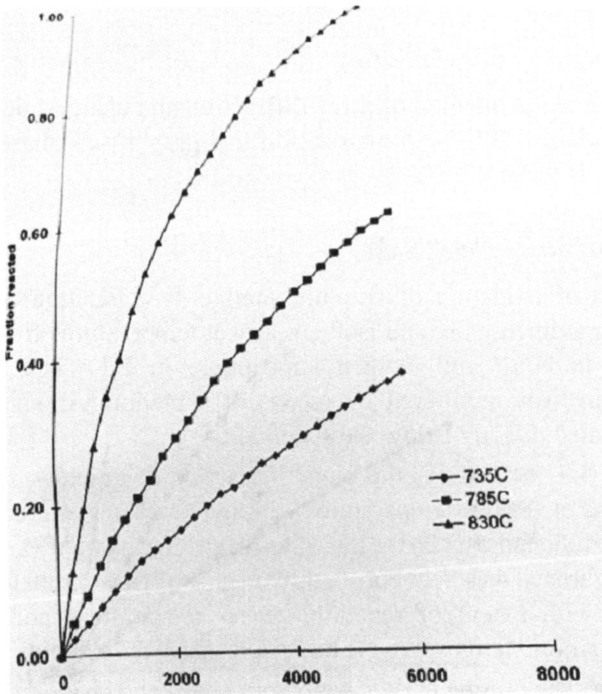

Fig. 5.10: Fraction Graphite Reacted as a Function of Time TG1 (oxygen)

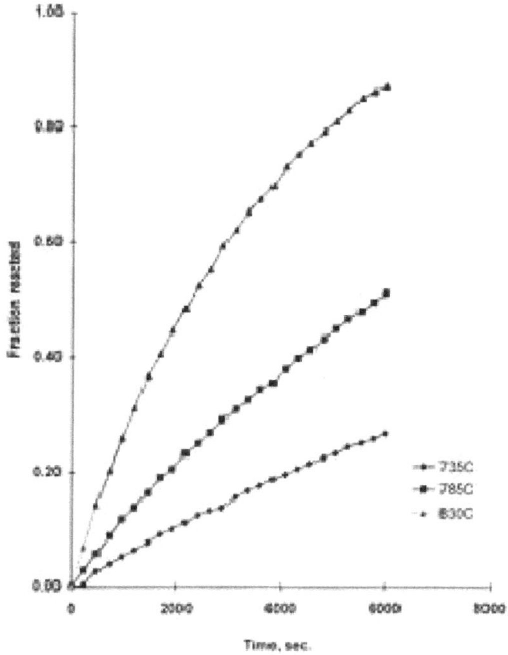

Fig. 5.11: Fraction Graphite Reacted as a Function of Time TG2 (air)

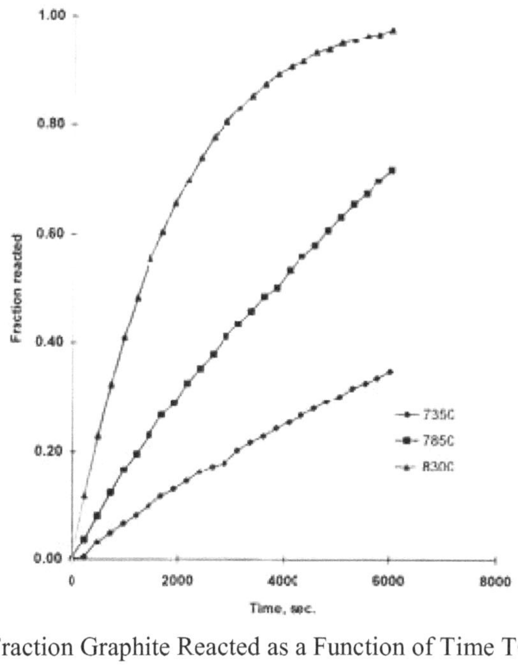

Fig. 5.12: Fraction Graphite Reacted as a Function of Time TG2 (oxygen)

The exact form of 'α' against 't' plot is determined by the geometry of the reaction interface. In most homogenous reactions, the whole range of decomposition fits one kinetic law but in heterogeneous reactions, one expression may hold for the initial few percent of the reaction to be followed by a second in the accelerating stage of the reaction and a third or fourth expression may be found to be describe the reaction in the decelerating stage of reaction.

5.3.3 *Summary of Observations*

- It has been found that grain size does not have impact on fraction decomposed for both size fractions. From the (α–t) curves, it is apparent that curves for two fractions in air and oxygen atmosphere are superimposed, which suggest that oxidation of graphite is independent of size fractions of (–30+150) BSS and such phenomenon has been reported by other authors.[85,88]
- Tamilnadu graphite used here contains higher amount of ash i.e., 4% and surface area is 1.4 m^2/gm.
- Degree of graphitsiation is very high i.e., 87% and percent of graphitised phase present in the sample is 93%.
- Ash contains impurities like quartz, feldspar, alkali-aluminium-iron-silicate hydrate, pyrite etc.
- Large amount of pores are present in the sample as depicted by SEM.
- It has very high surface area and large number of pores in the structure. Though degree of graphitisation is high and ash content is very low but its reactivity and activation energy is quite high.

CONCLUSION

- Graphite of three sources i.e. Palamu, Bolangir and Tamilnadu are used for refractory manufacture in India.
- Palamu graphite contains impurities like quartz, feldspar and potassium-iron-magnesium-aluminium silicates, hydrates. Bolangir graphite contains impurities like quartz, feldspar, corundum, non-troniteclay and calcium-magnesium-aluminium-silicates-maliss-clay. Tamilnadu graphite contains quartz, feldspar, pyrites and sodium-iron-aluminium-silicate hydrates.
- Ash percentage in Palamau graphite is 18%, Bolangir graphite 12% and Tamilnadu graphite is 4%.
- Surface area of Palamau graphite is 1 m^2/gm, Bolangir graphite is 0.4 m^2/gm and Tamilnadu graphite is 1.4 m^2/gm.
- Degree of graphitisation of Palamau graphite is 85%, Bolangir graphite is 98% and Tamilnadu graphite is 87%.
- Large amount of cracks and pores are present in all these samples as was clearly seen in SEM and impurities are mostly found around the pores and cracks in all the three graphite samples.
- Average equivalent diameter of the particle for Palamau graphite is 70 μm, for Bolangir graphite is 165 μm and for Tamilnadu graphite is 115 μm.
- It has been found that initial grain size does not have much impact on oxidation of graphite in air and oxygen atmosphere for all the three graphites. This has been confirm by the α-t curves. The plot of α-t curves are almost superimposed for both the size ranges in two atmospheres.
- These physico-chemical findings of the characteristics of three graphite samples would enable the manufacturers of refractory industry to select appropriate quality of graphite with proper size for manufacture of graphite containing refractory blocks.

REFERENCES

[1] Indian Mineral Book, Indian Bureau of Mines, Nagpur.

[2] Bailer, J.C., Emeleus, H.J., Nyholm, R. and Trotman, A.F., *Dickenson, Comprehensive Inorganic Chemistry*, Vol. 1, 1973, p. 1178.

[3] Nelson, J.B. and Riley, D.P., *Proc. Phys. Soc.* (London), 57 (1945) 477.

[4] Steward, E.G. and Cooke, B.P., *Nature*, 185 (1960) 78.

[5] Bacon, G.E., *Acta Crystl.*, 4(1951) 558.

[6] Gonser, B.W. and Hausuer, H.H., Modern Materials – Advances in development and application, Vol. 4, 1964.

[7] Cooper, C.F., *Interceram*, 32, 97, 1983.

[8] Boehm, H.P., Adv. *Catalysis*, 16 (1966) 179.

[9] Coulson, C.A., *Proc. 4th carbn conf.* (1959) 215.

[10] Ubbelohde, A.R., Nature, 180 (1957) 380.

[11] Amelinckx, S., Delavignette, P. and Heerschap, M., The Chemistry and Physics of carbon, Ed. P.L. Walker, Vol. 1, Dekker (1965).

[12] Nelson, J.B. and Riley, D.P., *Proc., Phys. Soc.*, 57 (1945) 477, 486.

[13] Pauling, L., The nature of chemical bond, 3rd ed., p. 235, Cornell University Press, Ithaca, N.Y. (1960).

[14] Bailer, J.C., Emeleus, H.J., Nyholm, R. and Dickenson, A.F., Comprehensive Inorganic chemistry, Vol. 1, p. 1179, Pergamon, Press.

[15] Ubbehode, A.R. and Lewis, F.A., "Graphite and its crystal compounds", Oxford Univ. Press, London, 1960.

[16] Nightangle, R.E., "Nuclear Graphite", *Academic Press*, New York, 1962.

[17] Franklin, R.K., Crsytallite growth in graphitiziung and non-graphitizing carbons, *Proc., Royal Soc.*, A 209, 196–218 (1951).

[18] Solnczewski, J.C. and Weis, P.R., Band structure of graphite, *Phys. Rev.*, 109, 272–279 (Jan 15, 1958).

[19] Walker, P.L., Jr. Carbon—An old but new material, *Am. Sci.*, pp. 259–293 (June, 1962).

[20] Bernal, J.D., *Proc. Roy. Soc.*, A106, 749 (1924).

[21] Girifalco, L.A. and Lad, R.A., *J. Chem, Phys.*, 25, 693 (1950).

[22] Brennan, R.O., *J. Chem. Phys*; 20, 40 (1952).

[23] Lipson, H. and Stokes, A.R. *Proc. Roy. Soc.* A181, 101 (1942).

[24] Debye, P. and Scherrer, P., Physikal Z., 8, 291 (1917).

[25] Boehm, H.P. and Coughlin, R.W., Carbon, 2 (1964) 1.

[26] Mochide, I. – Role of carbon in composite refractories, Taikabutsu overseas, Vol. 8, No. 4.

[27] Franklin, R.E., *Proc. Soc.*, A 209, 196, 1951.

[28] Ubbllohde, A.R. and Lewis, F.A., "Graphite and its crystal compounds", Clarendon Press, oxford, 1960, p. 5.

[29] Mishra, V.N. and Standish, N., *13th CMMI Congress – Metallurgy*, Vol. 4, ed. by L.E. Fielding and A.R. Gordon, The Australasian Inst. of Min. Metall, Australia (1986), 143.

[30] Beesting, M., Hartwell, R.R. and Wilknison, H.C., *Fuel*, 56(1977), 319.

[31] Alam, M. and Debroy, T., Metall,. *Trans. B*, 17B (1976), 565.

[32] Rao, Y.K. and Jalan, B.P., *Metall. Trans.*, 3 (1972), 2465.

[33] Turkdogan, E.T., Olsson, R.G. and Vinters, J.V., *Carbon*, 8 (1970), 545.

[34] Turkdogan, E.T. and Vinters, J.V., *carbon*, 8 (1970), 39.

[35] Turkdogan, E.T. and Vinters, J.V., *carbon*, 6 (1968), 467.

[36] Turkdogan, E.T. and Vinters, J.V., *carbon*, 7 (1969), 101.

[37] Tien, R.H. and Turkdogan, E.T., *carbon*, 8 (1970), 607.

[38] Tyler, R.J. and Smith, I.W., Fuel, 54 (1975), 99.

[39] Aderibigbe, D.A. and Szekely, J., *Ironmaking steelmaking*, 1 (1981), 11.

[40] Dutta, S. and Wen, C.Y., Ind. Eng. *Chem, Process Des. Dev*, 16 (1977), 20.

[41] Dutta, S. and Wen, C.Y., *Ind. Eng. Chem., Process Des Dev*, 16 (1977), 31/36.

[42] Adschiri, T., Shiraha, T., Ogawa, K. and Furusawa, T., *Proc Int. Conf. Coal Science*, (1985), 289.

[43] Adschiri, T. and Furusawa, T., *Fuel*, 65 (1986), 927.

[44] Bandopadhyay, D., Chakraborti, N. and Ghosh, A., *Steel Res.*, 62 (1991), 143.

[45] Roy, J.F. Boero, Carbon, 25 (1987), 477.

[46] Rodriguez, N.M., Marsh, H., Heintz, E.A., Sherwood, R.D. and Baker, R.T.K., *carbon*, 25 (1987), 629.

[47] De, A. Koranyi, *Carbon*, 27 (1989), 55.

[48] Ergun, S., *J. Phys. chem.*, 60 (1956), 480.

[49] RDCIS-NML Jt. project report on "Increasing oxidation resistance of carbon/graphite products by high temperature heat treatment", Report No. 2384, March 1995.

[50] Bandopadhyay, D., Ph.D. Thesis, IIT Kanpur, 1989.
[51] Langmuir, I., Trans Faraday Soc., 1921, Vol. 17, p. 621.
[52] Feldman, M.H., Goddel, W.V., Dienes, G.J. and Gossen, W., *Jr. of Appl. Phy*, 1952, Vol. 23, p. 1200.
[53] Kanter, M.A., *Phy. Rev*, 1957, Vol. 107, p. 655.
[53a] Rakshit, P.C., Physical Chemistry.`
[54] Warczok, A. and Utigard, T., *Steel Research,* 71, 8 (2000).
[55] Yagi, S. and Kunii, D., *Proc. 5th Int. Symp. On Combustion*, 1955, 231.
[56] Spitzer, R.H., Manning, F.S. and Philbrook, W.O., *Trans. Met. Soc.* AIME, 236 (1966), 726.
[57] Shen, J. and Smith, J.M., I and E.C. Fundamentals, 4 (1965), 293.
[58] Tartarelli, R., Morelli, F. and Capovani, N., Can. *J. Chem. Engg.*, 47 (1969), 201.
[59] Mckewan, W.M. and Smith, N.D., *Blast Furnace, Coke Oven and Raw Material Proc. AIME* 221 (1961), 3.
[60] Mathur, R., Rao, M.S. and Ghosh, A., Trans. *Indian Inst. Mat.*, 30 (1977), 45.
[61] Evans, J.W., Ph.D. Dessertation, State University of New York at Buffalo, 1970.
[62] Turkdogan, E.T., Iron Making Conference Proceeding, AIME, Chicago meeting 31 (1972), 438.
[63] Ausman, J.M. and Watson, C.C., *Chem. Eng. Sci.*, 17(1962), 323.
[64] Roy, A., Ph.D. Thesis, IIT Khargpur, Aug. 1988.
[65] Szekely, J. and Evans, J.W., *Met. Trans.*, 2 (1971), 1691 and 1699.
[66] Szekely, J. and Evans, J.W., *Chem., Eng. Sci.*, 26 (1971), 1901.
[67] Sohn, H.Y. and Szekely, J., *Chem. Engg. Sci.* 27 (1972), 763.
[68] Tien, R.H. and Turkdogan, E.T., *Met. Trans.*, 3 (1972), 2039.
[69] Papanastassion, D. and Bitsianes, G., *Met.Trans*, 4 (1973), 477.
[70] Szekely, J. and M.Propster, chem. *Eng. Sci.*, 30 (1975), 1049.
[71] Turkdogan, E.T., Olson, R.G. and Vinters, J.V.; *Carbon*, 8 (1970), 545.
[72] Koranyi, A.De; *Carbon*, 27 (1989), 55.
[73] Tien, R.H. and Turkdogan, E.T., *Carbon*, 8 (1970), 607.
[74] Adschiri, T., Shiraha, T., Ogawa, K. and Furusaw, T., *Proc. Int. Conf. Coal sci* (1985), 289.
[75] Inoue, H., Kiritani, Y. and Takahasi, Y., *Bull. Res. Inst. Miner. Dressing Metall.*, 31(1975), 127.
[76] Gokarn, A.N. and Doraiswamy, L.K., *Chem Eng. Sci.*, 3 (1973), 401.
[77] Sohn, H.Y. and Szekely, J., *Chem Eng. Sci.*, 27 (1972), 763.

[78] Ergn, S., Ind. Eng. Chem., 1955, Vol. 47, p. 2075.
[79] Roseberg, M. and Wicke, E., *chem. Ing. Tech*, 1956, Vol. 28, p. 181.
[80] Wicke, E., Chemical Reaction Engineering, *Ist European Symposium, Pergamon Press, Inc.*, N.Y., 1957, p. 61.
[81] Walker, P.L. Jr., Rusinko, F, Jr. and Austin, L.G., "Advan. in Catal"., *Academic Press, Inc.*, N.Y. 1959, Vol. 11, p. 134–221.
[82] Turkdogan, E.T., Koump, V., Vinters, J.V. and Perzak, T.F., *carbon* 1968, Vol. 6, p. 467/84.
[83] Rao, Y.K. and Jalan, B.P., *Met. Trans.*, 1972, Vol. 3, p. 2465.
[84] Standish, N. and Tanjung, A.F.A., *Fuel*, 1988, Vol. 67(5), p. 666.
[85] Ergun, S., U.S. Bur. *Mines Bull.*, 1962, Vol. 598, p. 38.
[86] Hippo, E. and Walker, P.L., *Fuel*, 1975, Vol. 54, p. 245.
[87] Tien, R.H. and Turkdogan, E.T., *carbon*, 1970, Vol. 8, p. 607.
[88] Van, K.H. Heek, Juntger, H. and Peter, W., *J. Inst. Fuel*, 1973, Vol. 46, p. 249.
[89] Rideal, E.K., Proc., *cambridge Phil. Soc.*, 1989, Vol. 35, p. 130.
[90] Sears, J.T., Muralidhare, H. and Wen, C.Y., Ind. Eng. *Chem. Proc. Des. Dev.*, 1980, Vol. 19, p. 358.
[91] Turkdogan, E.T., Olsson, R.G. and Vinters, J.V., *carbon*, 1970, Vol. 8, p. 545.
[92] Wu, P.C., Lower, W.E. and Hottel, H.C., *Fuel*, 1988, Vol. 67(2), p. 205.
[93] Johnson, J.L., Kinetics of coal gasification, John wiley and Sons, N.Y., 1979, p. 131.
[94] Alam, M. and Debroy, T., *Met. Trans. B.*, 1986, Vol. 17(3), p. 565.
[95] Szekely, J. and Aderibigbe, D.A.
[96] Agarwal, A.K. and Sears, J.T., *Ind. Eng. Chem. Process Des. Dev.* 1980, Vol. 19, No. 3, p. 365.
[97] Letrot, M. and Martin, G., Bull. Soc. *Chim. (*France), 1947, Vol. 14, p. 400.
[98] Walker, P.L. Jr., Shelefand, M., Anderson, R.A., *Chemistry and Physics of carbon,* edited by P.L. Walker Jr., Vol. 4, N.Y., 1968.
[99] Neumann, B., Kroger, C. and Fingas, E., *Z. Anorg. Chem.,* 1931, Vol. 197, p. 32165.
[100] Milner, G., Spivey, E. and Cobb, G.W., *J. Chem. Soc.*, 1943, p. 578.
[101] Vastola and Walker, P.L., Jr., *Jr. Chim. Phys*, 1961, Vol. 58, p. 20.
[102] Long, F.J. and Sykes, K.W., *Jr. Chin. Phys*, 1950, Vol. 47, p. 361.
[103] Long, F.J. and Sykes, K.W., *Proc. Roy. Soc.* (London), 1952, Vol. A2145, p. 100.

[104] Turkdogan, E.T. and Vinters, J.V., *Carbon*, 1972, Vol. 10, p. 97.

[105] Mehrotra, S.P. and Sinha, V.K., *Trans. ISIJ*, 1983, Vol. 23, p. 723.

[106] Johnson, J.L., Kinetics of coal gasification, *John Wiley and Sons*, N.Y., 1979, p. 131.

[107] Otto, K., Lehman, C., Bartoslewicz, L. and Shelef, M., *Carbon*, 1982, Vol. 20 (3), 102, p. 243.

[108] Fruend, H., Prepr. Pap.- Am chem. Soc., *Div. Fuel chem.,* 1985, Vol. 30(1), p. 311.

[109] Fruend, H., Fuel, 1986, Vol. 65(1), p. 63.

[110] Wigmans, T., Tromp, P., Moulijn, J.A., *Carbon*, 1984, Vol. 22(3), p. 319.

[111] Cerfontain, M.B., Meijer, R. and Moulijn, J.A., *Proc-Int. Conf. Coal Sci.*, 1985, p. 277.

[112] Yokoyama, S., Miyahara, K., Tanaka, K. and Tashiro, J., Sekiyu Gakkaishi, 1983, Vol. 26(96), p. 455.

[113] Alam, M. and Debroy, T., Met. Rans. B., 1986, Vol. 17(3), p. 565.

[114] Saber, J.M., Falconer, J.L. and Brown, L.F., J. Chem. Soc., *Chem. Commun.*, 1987, Vol. 6., p. 445.

[115] Nicol, A.W., Physico chemical methods of Mineral Analysis Plenum Press, New York and London, 1975, p. 322.

[116] Poddar, J., Ph.D Thesis, Dept. of Physics, Bangladesh University of Eng. and Technology, June 1995.

[117] Keattch, C.J. and Dollimore, D., An introduction to thermo-gravimetry, whitefriars Press Ltd.; London, 1975, p. 60–79.

[118] Choudhary, S.K. and Ghosh, A., RDCIS – IIT, Kanpur, Joint project report on, Study of coal char reactivity with reference to direct reduction processes, June, 1987.

[119] Ozgen, O.S. and Rand, B., Brit, *Ceramic Trans and Jr.*, 84, 70–76, 1985.

[120] Zamalloa, M., Ma, D. and Utigard, T.A., *ISIJ International,* Vol. 35 (1995), No. 5.

[121] Li, X., Rigaud, M., Palco, S., *Jr. Am. Ceram. Soc.*, 78(4), 965–71, 1995.

[122] Ichikawa, K., Nishio, H. and Hoshiyama, Y., *Taikabutsu Overseas* 14(1), 13–19 (1994).